Infrared Fiber Optics

Infrared Fiber Optics

Paul Klocek
Texas Instruments Incorporated

George H. Sigel, Jr.
Rutgers University

Roy F. Potter, Series Editor
Western Washington University

Volume TT 2

SPIE Optical Engineering Press

A Publication of SPIE—The International Society for Optical Engineering
Bellingham, Washington USA

Library of Congress Cataloging-in-Publication Data

Klocek, Paul.
    Infrared Fiber Optics.

    (Tutorial texts in optical engineering ; v. TT 2)
    Includes bibliographical references.
    1. Fiber optics.   2. Infrared technology.   I. Sigel,
George H., 1940-   II. Title.   III. Series.
TA1800.K56 1989   621.36'92   89-10781
ISBN 0-8194-0229-X

Published by
SPIE—The International Society for Optical Engineering
P.O. Box 10
Bellingham, Washington 98227-0010

Copyright © 1989 The Society of Photo-Optical Instrumentation Engineers

All right reserved. No part of this publication may be reproduced or distributed in any form or by any means without written permission of the publisher.

Printed in the United States of America

Foreword

Tutor: to teach or guide, usually individually, in a specific subject

The aim of the Tutorial Texts series is to fulfill the essential role of a tutor for selected topics in optical science and engineering. The Tutorial Texts are based on the SPIE short course program. The attendance and the evaluations of those attending a course provide a measure of the degree of interest in the subject matter as well as a qualitative statement about the potential of the course curriculum and lecture notes developing into a book.

This new series has been undertaken with much enthusiasm on the part of the SPIE Optical Engineering Press. These volumes reflect the excitement of working with our short course instructors to translate a series of lecture notes into stand-alone texts of a tutorial nature. Indeed, the Tutorial Texts are intended to accomplish, in book format, what the instructor/author does in the course environment. The scope, content, and presentation level are essentially the same in course and book.

This Tutorial Text on Infrared Fiber Optics provides the reader with a current view of the field of infrared fibers along with tutorial material for understanding the possibilities for future applications.

Roy F. Potter
Western Washington University
1 August 1989

TABLE OF CONTENTS

Section	Title	Page
1	Introduction	1
1.1	Need for Infrared (IR) Fiber Optics	3
1.2	IR Versus Visible Fiber Optics	4
1.3	Fundamentals of IR Fiber Optics	8
1.3.1	Propagation	8
1.3.2	Attenuation	12
1.3.3	Dispersion	24
1.4	References	28
2	IR Materials and Fibers	31
2.1	Halide Glasses	33
2.1.1	Composition	33
2.1.2	Material Preparation	37
2.1.3	Physical Properties	40
2.1.4	Waveguide Fabrication	46
2.1.5	Waveguide Properties of Fluoride Fibers	52
2.1.6	Mechanical Properties of Fluoride Fibers	56
2.1.7	Durability and Toughness	58
2.2	Chalcogenide Glasses	61
2.2.1	Composition	61
2.2.2	Materials Preparation	64
2.2.3	Physical Properties	66
2.2.4	Waveguide Fabrication	72
2.2.5	Waveguide Properties	77
2.3	Crystalline Materials	89
2.3.1	Composition	89
2.3.2	Materials Preparation	89
2.3.3	Physical Properties	90
2.3.4	Waveguide Fabrication	92
2.3.5	Waveguide Properties	98
2.4	IR Oxide Glasses	104
2.5	Hollow Waveguides	109
2.6	Summary	112
2.7	References	114
2.7.1	Subsection 2.1	114
2.7.2	Subsection 2.2	116
2.7.3	Subsection 2.3	118
2.7.4	Subsection 2.4	120
2.7.5	Subsection 2.5	120
2.7.6	Subsection 2.6	120
3	Applications of IR Fiber Optics	121
3.1	Overview	123
3.1.1	General Areas of Application for IR Fibers	123
3.1.2	IR Laser Propagation	123
3.1.3	IR Imaging	124
3.1.4	Active Components–Fiber Lasers	125
3.1.5	Sensors	125

TABLE OF CONTENTS (CONTINUED)

Section	Title	Page
3.1.6	Radiation-Hardened Links	125
3.1.7	Medical Uses	126
3.1.8	Nonlinear Optics	126
3.2	Telecommunications	127
3.3	IR Fiber Bundles Applications	131
3.3.1	Image Bundle	131
3.3.2	Tapered Bundle	135
3.3.3	IR Fiber Optic Reformatter	137
3.4	Single IR Fiber Applications	139
3.4.1	Temperature Sensor	139
3.4.2	Pressure Sensor	140
3.4.3	Remote Location of Detector or Source	141
3.5	References	146

LIST OF ILLUSTRATIONS

Figure	Title	Page
1	Attenuation Versus Wavelength for SiO_2	5
2	Attenuation Versus Wavelength for Oxides, Halides, and Chalcogenides	6
3	Diagram of Light Propagation in Stepped-Index, Graded-Index, and Single-Mode Optical Fibers	9
4	Hollow, Rectangular, Combination Metal-Dielectric Waveguide, TE_{10}	11
5	TE_{10} Mode in Hollow, Cylindrical, Metal Waveguide	11
6	Optical Attenuation Versus Wavelength	13
7	Band-Gap Absorption	15
8	Absorption Plots for Various Materials Classes	16
9	Water Absorption	17
10	Rare-Earth and Transition-Metal Absorption	18
11	V-Plot	21
12	Optical Fiber Attenuation Mechanisms	22
13	Intramodal Dispersion, Multimode Fibers Only	25
14	Intramodal Dispersion, Multimode and Single-Mode Fibers	25
15	Thermal Analysis	36
16	Fluoride Glass Fiber Materials, Thermal Analysis	37
17	Spectral Attenuation of Fluoride Glass Fiber (NTT)	38
18	Effect of REDOX Melting Conditions on the Optical Absorption in Fluoride Glasses	39
19	OH Absorption	39
20	Comparison of Optical Transmission of Fluoride Glasses with Silica	41
21	Scattering Loss Spectra for Bulk Fluoride Glasses	41
22	Light Scattering in Fluoride Glass and Silica	42
23	Refractive Index Spectra of Fluorozirconate Glasses	42
24	IR Edge for Various Fluoride Glass Compositions and Fiber	43
25	Material Dispersion of Fluoride Glasses	43
26	Absorption Spectra of Irradiated Fluorozirconate Glass	44
27	Fiber Drawing	47
28	Adding Core Glass To Create Solid Preform	49
29	Rotational Casting Process	49
30	Index of Refraction Across Preform	50
31	Reactive Vapor Transport Fluoride Glass Fiber	50
32	Refractive Index Versus Tube Wall Thickness for RVT Fibers	51
33	Localized Heat Zone Furnace for Drawing Fluoride Glass Fibers	52
34	Attenuation of Fluoride Glass Fibers	53
35	Loss Spectrum of a Fluoride Glass Fiber Produced at NRL	53
36	Transmission Loss Spectra for Step-Index Multimode Fibers	54
37	Comparison of Losses for Various Fluoride Glass Fibers	54
38	Breakdown of Total Loss in Fluoride Fiber	55
39	Measured Scatter Loss of Fluoride Fiber	55
40	IR Fiber Spectral Loss Measurement System	56
41	Strength of Fluoride Glass Fibers	57
42	Strength of ZBLA Fibers in Water	57
43	Leach Rates of ZBLA Glass with Pyrex	59
44	Water Absorption of Coated Fluoride Glasses	60
45	Glass-Forming Compositions of GeSbSe	62
46	Glass-Forming Composition Region of Systems SiSbS, SiSbSe, and GeAsSe	62

LIST OF ILLUSTRATIONS (CONTINUED)

Figure	Title	Page
47	GeAsTe Composition Diagram	63
48	Composition Diagram for GePS Glass System	63
49	Fabrication of Chalcogenide Glasses	65
50	Distillation Process Schematic Diagram	65
51	IR Transmissions of Purified Sulfide Glasses in Atomics Percent (Dashed curves show impurity bands)	66
52	IR Spectra of As_2Se_3 (R = reflection coefficient)	67
53	IR Spectra of As_2SeTe_2 (R = reflection coefficient)	67
54	IR Spectra of $Ge_{10}As_{50}Te_{40}$ (R = reflection coefficient)	68
55	IR Spectra of $Ge_{15}As_{10}Se_{75}$ (R = reflection coefficient)	68
56	IR Spectra of $Ge_{28}Sb_{12}Se_{60}$ and $Ge_{25}Se_{75}$	69
57	Optical Transmittance of 2-mm-Thick Disk of GeSeTe Glass, Showing (Solid Curve) Removal of Absorption Band at 13.0 μm by Heating Glass in Hydrogen	69
58	IR Transmission of Some GePTe Glasses Atomic Ratio	70
59	Apparatus for Fiber Drawing and Typical Data for Temperature Gradient at Neck-Down Region of Glass Rod	73
60	Preforms: (A) Square Extruded (B) Cast Cylindrical	73
61	Various Fibers	74
62	Drawing Apparatus of Chalcogenide Glass Fibers	75
63	Pyrex Glass Double Crucible Assembly for Preparing Arsenic-Sulfur Glass Fibers	76
64	Fiber Pulling Setup	76
65	Theoretical Attenuation in GeS Glasses	77
66	Spectral Dependence of Material Dispersion M(λ) of Glassy As_2S_3 and Region of Minimum Optical Losses in Glassy As_2S_3 (1) and As_2Se_3 (2)	78
67	Material Dispersion Versus Wavelength for $Ge_{28}Sb_{12}Se_{60}$	78
68	Intrinsic Attenuation Coefficient Versus Wavelength for $Ge_{25}Se_{75}$ and $Ge_{28}Sb_{12}Se_{60}$	79
69	Chalcogenide Glass Fiber Attenuation	80
70	Chalcogenide Glass Fiber Attenuation	82
71	IR Fiber Absorption Versus Wavelength	83
72	Transmission Loss Spectrum for $Ge_{20}Se_{80}$ Chalcogenide Glass Fiber	83
73	Chalcogenide Glass Fiber Attenuation (ϕ = diameter)	84
74	Transmission Loss Spectrum Around 10.6 μm for a $Ge_{22}Se_{20}Te_{58}$ Glass Optical Fiber	85
75	Spectra of the Optical Losses in Fibers Made of Chalcogenide Glasses	85
76	Drawing Process for Chalcogenide Fiber Bundles	86
77	Cross-Sectional Picture of IR Fiber Bundle, 2.0 mm in Diameter Including 200 AsS Glass Fiber Cores, Each 90 μm in Diameter	87
78	Coherent Image Bundle, 25-μm-Diameter Chalcogenide Glass Fibers, Shown in a Transmission IR Microscope	87
79	IR Fiber Optic Prototypes	88
80	Crystalline Fiber Extrusion	92
81	Schematic of Fiber Crystal-Growing Apparatus	93
82	Schematic Diagram of Laser-Heated Pedestal Growth System	93
83	Crystal Fiber Growth	94
84	Extruded KCl, 0.25 Inch in Diameter, 8×	96
85	KRS-5 Fiber From Laser Pedestal Growth Method	97

LIST OF ILLUSTRATIONS (CONTINUED)

Figure	Title	Page
86	Demonstration of Flexibility of a 150-μm C-Axis Sapphire Fiber (Ruler is marked in inches.)	97
87	Projected Transmission in IR Fibers	98
88	Projected Transmission in Crystalline IR Fibers	99
89	Projected Transmission Loss in IR Fibers	99
90	IR Loss Spectra for CaBr Fibers Grown in Different Atmospheres	100
91	Total Loss Spectrum of KRS-5 Fiber Measured by Cut-Back Method	101
92	Extruded Crystalline Fiber of KRS-5/KRS-6 Core/Clad	102
93	Refractive Index Variation with Wavelength for Five Glasses of Table 29	105
94	IR Transmission Spectrum of Germanate Glasses Prepared by Crucible Method and by VAD Method	105
95	Schematics of Porous Preform Preparation and Sintering Method	106
96	Calculated Theoretical Loss Spectrum of Germanate Glass Fiber	106
97	Measured Loss Spectrum of Germanate Glass Fibers with Silicone Resin Coating (Dashed Line) and without Any Coating (Solid Line)	107
98	Complex Index and Reflection Coefficient for SiO_2 and Pb-Glass	108
99	Experimental Result for Transmission of Pb-Glass Hollow-Core Fiber for CO_2 Laser Light (940 cm^{-1}) (Inner diameter is 1.0 mm; measured loss is for multimode transmission)	109
100	Transmission of Hollow Metallic Waveguide of Cylindrical Cross Section as Function of Inverse Bend Radius	110
101	Experimental and Theoretical Bending Losses of Nickel Cylindrical Waveguides Uncoated and Coated by Ge and PbF_2 Where X and O Correspond to Measured $E_\parallel$ and $E_\perp$, Respectively	110
102	Experimental Arrangement for Measuring 2.7-μm Power Transmission Properties (X-Y plotter and minicomputer were used for laser printer)	130
103	Coherent IR Fiber Optic Bundles	132
104	IR Fiber Optic Bundle Allowing Simultaneous Imaging at Two Different Wavelength Regions	133
105	Flexible Coherent Image Bundle (Detectors and cryogenics can be remoted, allowing more unobstructed image)	133
106	IR Image Bundle Configurations	134
107	IR Fiber Bundle, Field Flattener	134
108	Taper IR Fiber Bundle	135
109	Nonuniform Magnifying IR Bundle	136
110	IR Fiber Optic Image Reformation (Coherent)	137
111	Operation of Spectrometer Having IR Fiber Optic Reformatter	138
112	Temperature Sensor Diagram	139
113	Simplified Diagram of Pressure Sensor	141
114	IR Threat-Warning Receiver Diagram	142
115	CO_2 Laser Guide	144
116	FLIR Boresighting	145
117	Remote CO_2 Gas Analyzer/Remote Radiometry	145

LIST OF TABLES

Table	Title	Page
1	Comparison of Silica-Based Versus IR Optical Material Compositions for Fiber Optics	5
2	General Comparison Between Silica-Based and IR Fiber Optics Technologies	6
3	Comparison Between Silica-Based and IR Fiber Optics in Terms of Attenuation, Dispersion and Propagation	7
4	Summary of Specific Absorption Coefficients	17
5	Wavelengths at Which Vibrational Absorptions of Various Impurities Are Active	19
6	Typical Heavy Metal Fluoride Glasses Used in the Fabrication of IR Fibers	34
7	Glass-Forming Systems Not Containing ZrF_4 or HfF_4 and Guide to Available Data	35
8	Compositions of Heavy-Metal Fluoride Glasses and Their Acronyms as Used in the Text	36
9	Purification Requirements in Fluoride Glasses Impurity Levels (PPB) Causing 0.01 dB/km Loss	38
10	Optical Properties of Fluoride Glasses	44
11	Mechanical Properties of Fluorizirconate Glasses	45
12	Elastic Properties of Fluorides Compared With Oxides	46
13	Physical and Thermal Properties of Fluorides	46
14	Promising Methods for Further Toughening Fluoride Glasses	60
15	Chalcogenide Glass Compositions; Various Atomic Ratios	61
16	Absorption Coefficients in Bulk Chalcogenide Glasses	70
17	Physical Properties of Chalcogenide Glasses	71
18	Comparison of Attenuation Minima and Material Dispersion Zeros in Chalcogenide Glasses	79
19	Laser Power Transfer in Chalcogenide Glass Optical Fibers	86
20	Strength of Chalcogenide Glass Optical Fibers	86
21	Materials for Infrared Fibers	89
22	IR Transmission in Crystalline Materials for Fibers	90
23	Absorption Coefficients of IR Crystalline Materials for Fibers	91
24	Physical Properties of IR Crystalline Materials for Fibers	91
25	Fabrication Parameters of IR Crystalline Fibers	95
26	Material Dispersion Zeros and Theoretical Attenuation Minimum in Crystalline IR Fibers	101
27	Attenuation in Crystalline IR Fibers	103
28	Composition and Some Properties of IR Oxide Glasses	104
29	Optical Properties of Glasses in Table 28	107
30	Hollow Waveguide Transmission	111
31	Summary of IR Optical Fibers	113
32	Applications of Infrared Fibers	124
33	Most Promising Candidates for Ultralong Link, Underwater Mid-IR Fiber Optics Systems	128
34	Proposed 2- to 5-μm Semiconductor Lasers: Important Features	128
35	Performance of Color Center Lasers	129
36	Potential Detectors for Mid-IR Fiber Optics	129
37	Criteria for Fiber Optic Image Bundle	131

Preface

We first taught the SPIE short course from which this book is adapted in May 1984. We agreed to teach the course for two reasons: first, to explain and demonstrate the potential of IR fiber technology and to encourage its application, and second, because it afforded us the opportunity to pull together and review the many different infrared fiber optic technologies.

Due to the diversity of applications and the many different IR fiber technologies under development, this Tutorial Text necessarily covers a lot of material. A general review of the fundamental principles of propagation, attenuation, and dispersion as they relate to dielectric and hollow waveguides made of oxide, halide, or chalcogenide glasses and metals is the starting point. This is intended to allow the reader to move through the chapters on each fiber technology and its applications recognizing its advantages, disadvantages, the relative state of the art, and its future potential. The book moves from the fundamentals through materials, fabrication, physical properties, and finally applications.

As an adaptation from a short course, the book gives a complete overview of the entire field but often with brevity. The general level of the book is intended to allow someone new to the field to gain a useful insight, as well as someone established in the field to have access to a comprehensive review of the different IR fiber optic technologies.

Acknowledgments

The authors would like to acknowledge the financial support of Texas Instruments Incorporated for the manuscript preparation, the Technical Publications Department of Texas Instruments for preparing the manuscript, Gloria Beduhn for editorial support, Dr. James M. Florence for technical assistance, and Annette Woktowicz for general assistance.

Paul Klocek August 1989
Texas Instruments Incorporated

George H. Sigel, Jr.
Rutgers University

Section 1
Introduction

1.1 NEED FOR INFRARED (IR) FIBER OPTICS

Demands for technological advances in the areas of communications, medicine, the military, and industry are constantly increasing. Long repeaterless communication links, remote optical powering, flexible laser surgery, remote sensing and imaging, and a greater degree of freedom in electro-optical system design are some of these desired advances. Infrared (IR) fiber optics offers an answer to many of these needs. Development of ultra-low-loss communication links is possible with IR fiber optics, as is remote optical powering. CO_2 and other IR laser outputs can be flexibly guided by IR fibers. IR fibers and bundles also offer a new degree of freedom in system design such that remote thermal sensing and imaging are possible, along with increased system capabilities and functional design. They also offer a variety of novel fiber optic sensors for various temperature, pressure, and chemical environments, and other applications in military, medical, commercial, and industrial settings. IR fiber optics is a new technology that will both impact many of today's technological needs and provide a basis for future needs.

1.2 IR VERSUS VISIBLE FIBER OPTICS

Silica-based fiber optics is a well-developed technology that has had major impact on telecommunications, medicine, and industry. Optical fibers fabricated with silica-based glass have achieved the intrinsic attenuation limits of 0.2 dB/km at 1.5 μm.[1] Widespread production of optical fibers proves their dominant role in communications for both voice and data; in sensors for medicine, industry, and the military; and in optical and electro-optical systems in a myriad of fields. The general advantages of waveguides are well known not only for the visible and near-IR portions of the spectrum that the silica fibers transmit but also for other parts of the spectrum such as the radar regime, where hollow metallic waveguides are used. One part of the electromagnetic spectrum that is of widespread interest is the mid-IR. The many IR electro-optical systems whose design could be greatly enhanced or made possible with the inherent flexibility of fiber optics and the possibility of ultra-low-loss telecommunications have driven the development of the topic of this book, *Infrared Fiber Optics*.[2,3,4,5,6] Specifically, this book addresses the IR fiber technologies that operate in the 2- to 14-μm spectral region.

Silica optical fibers are limited in the wavelength they can transmit to approximately 3 μm. IR fiber optics that transmit from 2 to 14 μm require entirely different materials for their fabrication. Fortunately, some of these IR fiber materials can take advantage of the existing fabrication technology developed for silica fiber optics.

The fundamentals that describe the operation of silica optical fibers in terms of attenuation, propagation and dispersion are not changed when considering IR optical fibers with the exception of the hollow waveguides. Table 1 is a comparison of silica-based materials and some of the many IR optical materials from which optical fibers have been fabricated. Figure 1 illustrates the wavelength region in which silica-based fiber optics operate. For comparison, Figure 2 illustrates the operating wavelengths for many IR fiber optic materials. Table 2 presents a general comparison between silica-based and IR fiber optics technology. Table 3 compares the two technologies in terms of propagation, attenuation, and dispersion.

**TABLE 1. COMPARISON OF SILICA-BASED VERSUS
IR OPTICAL MATERIAL COMPOSITIONS FOR FIBER OPTICS**

Silica-Based	Infrared	
SiO_2	**Glass**	**Crystalline**
SiO_2-B_2O_3	ZrF_4-AlF_4-BaF_2-LaF_3	CsI
SiO_2-BaO	PbF_2-MnF_2-CrF_3	CsBr
SiO_2-PbO	BaF_2-LaF_3-ThF_3	KCl
SiO_2-TiO_2	BeF_2	KRS–5 (ThBrI)
SiO_2-P_2O_5	ZnCl	AgCl
SiO_2-Na_2O-CaO	Ge-Se	AgBr
Other modifiers (K_2O, MgO and formers: (Al_2O_3))	Ge-S	ZnSe
	Ge-Sb-Se	**Hollow**
	As-S	Metallic
	As-Se	Dielectric
	Ge-As-Te	
	GeO_2-BiO-TlO	

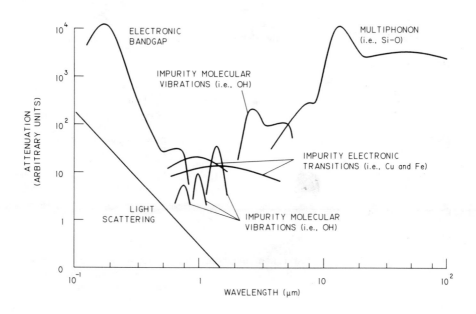

Figure 1. Attenuation Versus Wavelength for SiO_2

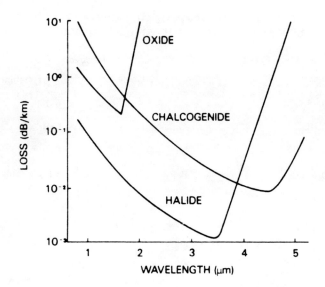

Figure 2. Attenuation Versus Wavelength for Oxides, Halides, and Chalcogenides (Reference 8)

TABLE 2. GENERAL COMPARISON BETWEEN SILICA-BASED AND IR FIBER OPTICS TECHNOLOGIES

	Fiber Optics	
Property	Silica-Based	Infrared
Transmission range	0.2 to 4.0 μm	0.2 to 15.0 μm
Material types	Silica, silicate, germanate, borosilicate glasses; polymers such as polystrene	Halide crystals and glasses, semiconductor crystals and glasses, heavy metal oxides and metals
Mechanical properties	Very strong	Moderate, intrinsically lower than SiO_2
Thermal properties	Very high temperature (> 1,000°C)	Moderate (~300 to 500°C) to very low (< 100°C)
Chemical properties	Very stable and durable	Moderate to hygroscopic
Level of technology	Production	Development
Problems to be solved	Improved economical production	Materials development in purification and durability, fiber, strength, cabling and production

TABLE 3. COMPARISON BETWEEN SILICA-BASED AND IR FIBER OPTICS IN TERMS OF ATTENUATION, DISPERSION AND PROPAGATION

Property	Silica-Based	Infrared
Theoretical attenuation minimum	0.2 dB/km at 1.55 μm	10^{-2} to 10^{-4} dB/km at 2.5 to 8 μm
Achieved theoretical minimum	Yes	No
Material dispersion zero	1.3 μm	1.5 to 8 μm
Delay distortion (intra and intermodal)	Low ($<$ 10 ps/nm $\cdot$ km)	Low
Wavelength difference between theoretical minimum and dispersion zero	Small	Often larger than SiO_2
Single-mode core size	6 μm	10 to 25 μm

1.3 FUNDAMENTALS OF IR FIBER OPTICS

This review of the fundamentals of fiber optics applies to all types of optical fibers, including IR fiber optics. The principles of light propagation, attenuation, and dispersion are discussed. Optical materials useful for IR fiber optics should possess the following properties: low material dispersion, high energy band-gap, λ^{-4}-dependent light scattering, long wavelength multiphonon edge, mechanical properties that result in high strength, and thermal and chemical stability. The fundamentals with regard to hollow waveguides differ from dielectric waveguides and are reviewed under separate headings.

A more thorough description of the general fundamentals of fiber optics can be found in many references on the subject, including Reference 9.

1.3.1 Propagation

1.3.1.1 Dielectric Fibers

The principle of total internal reflection is a ray method of describing light propagation in all dielectric optical fibers, whether silica-based or IR fiber. However, the modal method, which is derived from solutions to Maxwell's equations with the imposed boundary conditions of the waveguide, is required to explain all the properties of a waveguide, such as single-mode behavior. By considering the optical fiber shown in Figure 3, many of the basic propagation properties can be defined. The numerical aperture, NA, is given by:

$$NA = (n_1^2 - n_2^2)^{1/2} \qquad (1)$$

or

$$NA = \sin\theta = n_1 \sin\theta' \qquad (2)$$

where

n_1 = refractive index of core
n_2 = refractive index of clad
$\theta' = \theta_c$
$\theta_c = \cos^{-1} n_2/n_1$ = critical angle for total internal reflection.

The NA defines the solid angle over which the fiber can collect light and guide it to the other end. The NA is limited to a value of 1.0 even if numerically larger values can be obtained because of large differences between n_1 and n_2, as this would violate Maxwell's equations. An optical fiber can operate without a clad

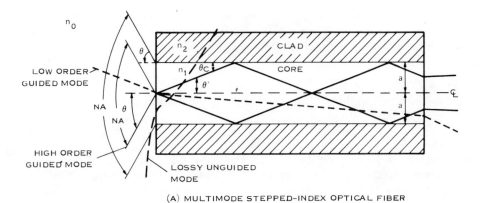

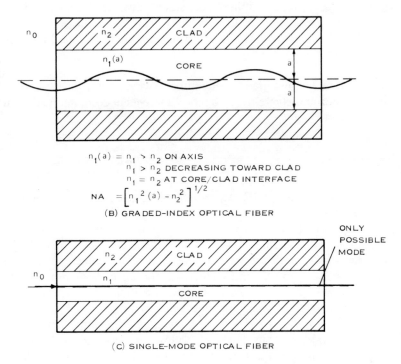

Figure 3. Diagram of Light Propagation in Stepped-Index,
Graded-Index, and Single-Mode Optical Fibers

provided $n_o < n_1$ but this generally results in greater attenuation or transmission loss through the fiber from surface effects and an NA equal to 1.0 resulting in lossy high-order modes. NA^2 equals the fraction of light collected by the fiber from a Lambertian source. Any light falling within the NA excites a guided mode within the fiber and outside the NA excites an unguided mode.

The normalized frequency, V, is given by:

$$V = 2\pi(a/\lambda)NA \qquad (3)$$

and the number of guided modes is $V^2/2$. A single-mode fiber occurs where V = 2.4. This is the result of all the guided modes being cut off except for the lowest order hybrid mode. With everything being equal, Equation (3) shows the effect of moving to longer wavelengths from silica-based fibers to IR fibers. Thus, single-mode IR fibers are larger in diameter than silica-based fibers, which is advantageous for coupling and splicing considerations. This does require, however, that IR bundle fibers be larger to reduce crosstalk, which lessens the bundle modulation transfer function (MTF) or resolution. Bundles and MTF are discussed in Subsection 3.3.

Figure 3 shows graded-index fibers, which carry fewer modes than stepped-index fibers, and single-mode fibers. Reference 9 contains further information on this subject.

1.3.1.2 Hollow Fibers

Another type of IR waveguide to consider is the hollow type shown in Figures 4 and 5. Propagation of electromagnetic energy in a hollow waveguide can be described by a ray method or a modal method by solving Maxwell's equations.[11] In general, cylindrical hollow waveguides propagate TE_{0n} (transverse electric) modes with the least attenuation while hollow rectangular waveguides propagate TE_{m0} modes most efficiently because these modes interact the least with the waveguide inner walls (grazing incidence reflection). TM_{m0} modes would propagate between the side walls of Figure 4 for the same reasons. The propagation of the energy can be viewed as reflection off the internal surface of the waveguide. Various types of hollow IR waveguides have been fabricated, including: cylindrical and rectangular metallic, cylindrical dielectric, and cylindrical and rectangular metallic with inner dielectric coatings. Whatever the hollow waveguide material, it must be highly reflective to the IR radiation of interest to be effective. For hollow dielectric waveguides, this generally means selecting a dielectric with an anomalous refractive index dispersion characteristic at or near the intended operating wavelength of the waveguide. This results in the real part of the refractive index with a value at or less than unity while the concomitant imaginary part (extinction coefficient) is usually a moderate value (1 to 10). Ideally, the dielectric material of a hollow waveguide should possess a real part as close to unity as possible and a small extinction coefficient.

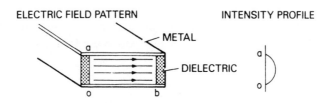

Figure 4. Hollow, Rectangular, Combination Metal-Dielectric Waveguide, TE$_{10}$ (Reference 10)

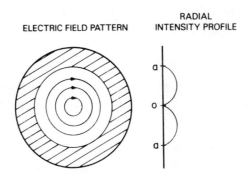

Figure 5. TE$_{10}$ Mode in Hollow, Cylindrical, Metal Waveguide (Reference 10)

For metal hollow waveguides, the metal material inherently provides high IR reflectivity but, unlike those made of dielectric material, they usually possess a large refractive index (i.e., aluminum, n = 25 + i67) at the wavelengths of interest. Like the dielectric, the hollow metal waveguide propagates energy most efficiently in low-order modes. This suggests maximizing the angle of incidence normal to the waveguide wall.

In contrast to solid dielectric optical fibers, some of the electromagnetic energy propagating in a hollow waveguide penetrates a lossy wall. As with dielectric fibers, but more so, the higher order modes in a hollow waveguide are more rapidly attenuated. The best choice for propagation in a hollow waveguide is low-order modes whose polarization is parallel to the wall as seen in Figures 4 and 5. This is achieved by optically coupling the desired mode or modes from a laser into the waveguide. Low-order modes can also be excited using incoherent black-body radiation coupled into the hollow waveguide at grazing incidence to the waveguide wall. This will result in higher attenuation than the selected polarized light from the laser because of increased interaction of the waveguide wall with some of the polarizations.

The NA of any hollow waveguide is essentially 1.0 since there are no "unguided" modes. Practically, however, a hollow waveguide would not be operated at other than small NAs to maintain some transmitted energy as previously discussed.

The physical dimensions of the waveguide will, however, limit the frequency or wavelength of light that can be guided. An example of this can be seen for the TE modes in a rectangular waveguide where the frequency cutoff is given by:[12]

$$\omega_{mn} = \frac{c\pi}{\sqrt{\mu_o \epsilon_o}} \left(\frac{m^2}{a^2} + \frac{n^2}{b^2}\right)^{1/2} \quad (4)$$

where

c = speed of light in vacuum
μ_o = magnetic permeability
ϵ_o = electric permittivity.

For a rectangular IR hollow waveguide (Figure 4) operating at 10.0 μm, "a" can equal 200 μm or greater and "b" can be anything as long as it is several times "a."

1.3.2 Attenuation

1.3.2.1 Dielectric Fibers

The total optical attenuation, α_T, in a dielectric or semiconductor material is given generally by:

$$\alpha_T = Ae^{(a/\lambda)} + Be^{-(b/\lambda)} + C/\lambda^4 + D(\lambda) + E(\lambda) + F/\lambda^{0-4} + G\lambda^{1.5-3.5}. \quad (5)$$

Figure 6 describes the various sources of attenuation and their wavelength dependence on optical attenuation. Attenuation in a material is usually given in cm^{-1}, whereas fiber attenuation is usually given in dB/km or dB/m,

$$\frac{dB}{m} = \frac{10 \log_{10}\left(\frac{power_{in}}{power_{out}}\right)}{length~(m)}.$$

The conversion is 2.3 $\times$ 10^{-3} cm^{-1} per dB/m.

The first term in Equation (5) [Ae$^{(a/\lambda)}$] is caused by intrinsic electronic absorption. Band-gap absorption occurs when incident light of sufficient photon energy (frequency ω) is absorbed by a valence electron that then moves across the forbidden band into the conduction band. The forbidden band defines the minimum photon energy required to promote a valence electron to the conduction band. The resulting absorption coefficient is exponentially dependent on the wavelength as in Equation (5).

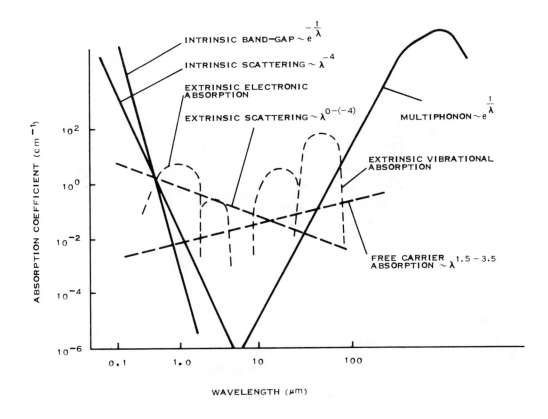

Figure 6. Optical Attenuation Versus Wavelength (Reference 13)

Intrinsic band-gap absorption, for which the electronic transition is from the valence band to a conduction band, generally has the greatest absorption coefficient of the electronic processes. Other intrinsic band-gap processes include transitions by direct or indirect means from the valence band to the lowest conduction band; absorption that creates excitons, which are bound electron-hole pairs; and free-carrier absorption with transitions from low to high conduction bands. These processes are mentioned in order of decreasing activation energy.

Free carrier absorption is the seventh term ($G\lambda^{1.5-3.5}$) in Equation (5). It is caused by electrons in the conduction band (or holes in the valence band) that are accelerated by the electromagnetic field in the material (IR wavelengths in the case of IR fibers) and scattered inelastically by phonons and ionized impurities, thus surrendering energy to the lattice of the material. This process can be described by the Drude model as:

$$\alpha_{fc} = \frac{Nq^2 \lambda^2}{m^* 8 \pi^2 n c^3 \tau} \tag{6}$$

where

N = carrier concentration
q = charge per carrier
λ = wavelength
m^* = reduced mass
n = index of refraction
c = speed of light in vacuum
τ = relaxation time.

A more realistic equation is:[14]

$$\alpha_{fc} = c_1 \lambda^{1.5} + c_2 \lambda^{2.5} + c_3 \lambda^{3.5} \tag{7}$$

where

c_1 = constant for acoustic phonon scattering
c_2 = constant for optical phonon scattering
c_3 = constant for ionized impurity scattering.

Equation (7) will result in free carrier absorption dependent on $\lambda^{1.5}$ to $\lambda^{3.5}$. The specific type of material will determine where it falls in this range, primarily depending on its carrier mobility.

Electronic absorption from extrinsic processes, the fourth term $[D(\lambda)]$ in Equation (5), can be significant. Point defects, structural imperfections such as dislocations and grain boundaries, and impurity atoms can create electrically active states within the forbidden band so that low energy absorption can take place. The point defects and structural imperfections are more important in crystals than in glasses. Impurity atoms are present in both glasses and crystals. Rare-earth and transition-metal ions are of greatest concern because their low-lying electronic configurations allow absorption at the short IR wavelengths. Figure 7 illustrates the electronic absorption processes in a flat-band and E-versus-k model.

The second term $[Be^{-(b/\lambda)}]$ in Equation (5) is absorption caused by the coupling of the electromagnetic energy to the dipoles created by the molecular or lattice vibrations of the material. This so-called multiphonon absorption is exponentially dependent on wavelength, as the absorption of an incident photon can occur by coupling to the fundamental phonons, their overtones, or combinations thereof. Because of the particular selection rules involved, the multiphonon edge is often not a smooth exponential but may have distinct features or absorption bands, each related to a particular longitudinal or transverse optical or acoustic phonon or combinations thereof.

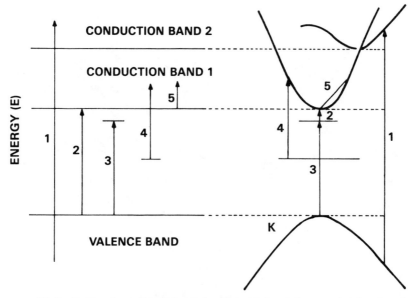

(1) Excitation from the valance band to higher lying conduction bands,
(2) Excitation across the band gap,
(3) Exciton formation,
(4) Excitation from imperfections,
(5) Free-carrier excitation

Figure 7. Band-Gap Absorption (Reference 15)

The fundamental phonon, often called the reststrahlen band, occurs at the frequency V_o given by:

$$V_o = \frac{1}{2\pi}\left(\frac{f}{\mu}\right)^{1/2} \tag{8}$$

where

f = force constant between the atoms of the dipole

μ = reduced mass of the atoms.

Equation (8) gives a simple explanation of why silica is limited to 3 μm and why IR fibers are made of different materials. Silica is composed of silicon and oxygen, both of which have relatively light atomic masses, and they are strongly bonded together by covalent bonds. Therefore, an IR material should be composed of heavy atoms to maximize the reduced mass, and be bonded weakly to minimize the force constant, resulting in a low-frequency, long-wavelength reststrahlen band and IR transparency. Figure 8 is a general plot for the multiphonon absorption edge in a variety of material families.

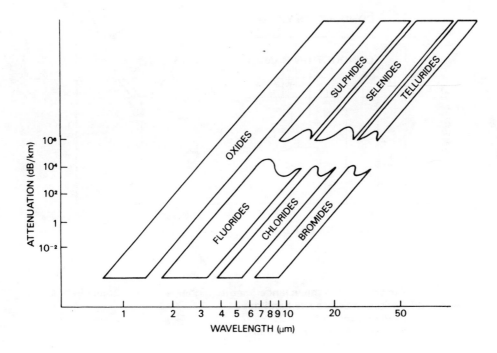

Figure 8. Absorption Plots for Various Materials Classes (Reference 16)

Multiphonon absorption is an intrinsic process and establishes the fundamental limit in the long wavelength attenuation of a material. An important extrinsic absorption process for IR optical fibers is impurities whose local vibrational modes [the fifth term in Equation (5), $E(\lambda)$] are IR active in the region of frequencies higher than the multiphonon edge. These absorptions may exclude regions of interest that appear transmissive from the theoretical V-curve, explained later in this subsection. Some of the most common vibrational impurities include OH, H_2O, and O_2. An example for H_2O can be seen in Figure 9, which shows the many active absorption bands. This demonstrates that a material containing H_2O may have more impurity vibration absorption bands than the commonly noticed one at 2.8 μm. Vibrational impurities are important not only as dissolved species in the optical material of the fiber but also as macroscopic heterogeneities, imperfections, and surface impurities.

The specific absorption coefficients for various extrinsic processes for both the electronic and vibrational types are useful for quantifying their effects on the light propagation in the material of the fiber. The specific absorption coefficients, having units of dB/km/ppm, relate the attenuation to the concentration of the impurity in the optical fiber. Some specific absorption coefficients for various types of impurities that absorb in the 2- to 6-μm region are listed in Table 4. For determining attenuation goals for an infrared optical fiber, use can be made of the simple expression B/c = specific absorption coefficient, where B is the attenuation coefficient goal of the fiber and c is the concentration of impurity. For

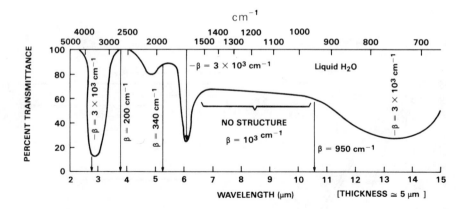

Figure 9. Water Absorption (Reference 17)

TABLE 4. SUMMARY OF SPECIFIC ABSORPTION COEFFICIENTS (REFERENCE 18)

Impurity	Specific Absorption Coefficient (dB/km/ppm) in the 2- to 6-μm Region
Clusters	Up to $\sim 5 \times 10^3$
Macroscopic imperfections	Up to $\sim 5 \times 10^3$
Rare-earth ions	Up to 40
Transition-metal ions	Negligible to 130
Ionic vibrational impurities	Up to 1×10^4 (for OH on-resonance)

example, if B = 10^{-3} dB/km at a frequency at which an impurity had a specific absorption coefficient of 10^4 dB/km/ppm, then c must be less than 10^{-7} ppm, which corresponds to a fractional concentration of 10^{-13}. If c = 1 ppm and B = 10^{-3} dB/km, the specific absorption coefficient must be 10^{-3} dB/km/ppm. Since the specific absorption coefficients are generally orders of magnitude higher than this value and concentrations of common impurities of 10^{-7} ppm are essentially impossible to achieve or even measure, B = 10^{-3} dB/km would thus be unobtainable at this frequency and would only be possible at a frequency away from any extrinsic process of absorption. Clearly, operation at frequencies where extrinsic absorption is active can be achieved only if higher values of B are tolerable.

Specific absorption coefficients as functions of wavelength for the rare-earth and transition metals are shown in Figure 10. Table 5 lists the wavelengths at which the vibrational absorptions of various impurities are active. There are many other possible impurities, depending on the material and fiber fabrication processes involved.

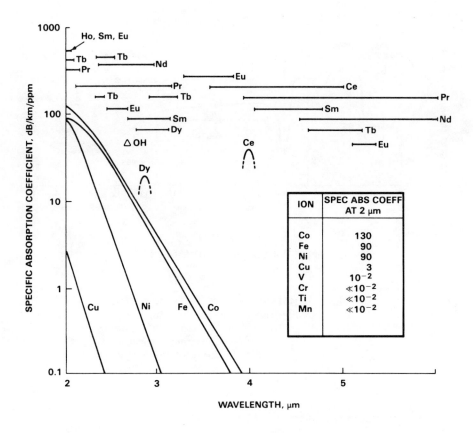

Figure 10. Rare-Earth and Transition-Metal Absorption (Reference 18)

The third term (C/λ^4) in Equation (5) is caused by light scattering in the material. This term is usually dominated by Rayleigh scattering but also includes Brillouin scattering, which results from inelastic interactions with propagating fluctuations in the dielectric function. Rayleigh scattering is caused by thermally arrested density fluctuations in the material, which lead to localized changes in the dielectric constant. These static on stationary fluctuations elastically scatter light in directions other than the direction of the incident light, but at the same wavelength. Brillouin scattering is caused by propagating fluctuations in the dielectric constant as a result of the intrinsic acoustic phonons in the material. Glasses have an additional contribution to Rayleigh scattering because of thermally arrested compositional fluctuations. Rayleigh and Brillouin scattering are proportional to ω^4; therefore, they affect the high frequency region of the optical spectrum (ultraviolet and visible) more than the low frequency IR region.

Scattering coefficients for density and compositional fluctuations are given respectively by:[19]

$$\alpha_d = \frac{8}{3} \frac{\pi^3}{\lambda^4} (n^8 p^2) \, kT \, (\beta_T \text{ or } \beta_S) \qquad (9)$$

where

   n = index of refraction

   p = photoelastic coefficient

   k = Boltzmann constant

   T = glass transition temperature for glasses or simply temperature for crystals

   β_T = isothermal compressibility for Rayleigh scattering

   β_S = adiabatic compressibility for Brillouin scattering

TABLE 5. WAVELENGTHS AT WHICH VIBRATIONAL ABSORPTIONS OF VARIOUS IMPURITIES ARE ACTIVE (REFERENCE 19)

Impurity (left)	Impurity (right)
SH^-	CO_2
$-CH_3$	CO
$-CH_2$	COF_2
Si-OH	OH^-
$COCl_2$	BH_4^-
Amides	NH, NH_2
H_2O liquid	$KHCO_3$
NH_4^+	N_2O
HCO_3^-	$KHCO_2$
KNO_3	P-OH
NO_2^-	NO_3^-
CN^-	$X=Y, X=Y=Z^*$
Si-H	Si-O-Si
NCO^-	O_3^-
N_3	BO_2^-
Aromatics	BO_3^-
Ge-O-Ge	Carbonyl
Esters	CO_2^-
Al-O-Al	CO_3^{2-}
C-Cl	P-O-C
B-N, B-O	SO_2
CF_2, CF_3	SO_4^{2-}
ClO_3	

*X,Y = C,N; Z = C,N,O,S: X=Y means C≡C, C≡N, etc.

Correlation chart showing the absorption lines between 2 and 10 μm of some chemical bonds, molecular ions, radicals, and compounds. Key: • peak positions; — range of peak positions in various materials.

and

$$\alpha_c = \frac{16\pi^3}{3\lambda^4} \left(\frac{dn}{dc}\right)^2 (\Delta_c)^2 \, \delta V \tag{10}$$

where

$(\Delta_c)^2$ = mean square concentration fluctuation

δV = volume of fluctuation.

Extrinsic scattering processes, the sixth term (F/λ^{0-4}) in Equation (5), give rise to attenuation greater than the intrinsic Rayleigh scattering. Extrinsic scattering arises from impurities and imperfections or heterogeneities in the material, where the scattering dependence on ω can range from ω^4 to ω^0. Mie scattering is an example for which the scattering scales with ω at some power less than 4, which increases the amount of scattering at lower frequencies. Examples of ω^2 scattering include scattering by varying strain fields in a polycrystalline or glass fiber or by regions of devitrification in glasses.

Additional attenuation mechanisms that are not included in Equation (5) are intensity-dependent Brillouin and Raman scattering. They result from inelastic interactions with propagating refractive index fluctuations where, generally speaking, those involving acoustic phonons are Brillouin and those involving optic phonons are Raman. Raman scattering involves either the incident light photon emitting a phonon and reradiating the remaining energy as another photon, or a phonon being absorbed as the incident photon is converted to the scattered photon. Brillouin scattering occurs when the incident photon is scattered by an acoustic phonon. By energy conservation, the scattered photon in either Raman or Brillouin scattering has a frequency shift on either side of the incident frequency. At low optical power, there is spontaneous Raman and Brillouin scattering. As the power is increased, stimulated Raman and Brillouin scattering can result in severe attenuation as energy is transferred to the stimulated Brillouin backward-traveling wave or to the Raman wave. It is the stimulated processes that define the limits of the power-handling capability of the optical fiber, assuming catastrophic damage has not occurred. Critical powers for these processes for a probable ultra-low-loss single-mode IR optical fiber where λ = 2.55 μm are given by:[21]

$$\text{Stimulated Brillouin:} \quad P_B = 21 \frac{A}{g_B L} \rightarrow 1.2 \text{ mW} \tag{11}$$

$$\text{Stimulated Raman:} \quad P_R = 16 \frac{A}{g_R L} \rightarrow 440 \text{ mW} \tag{12}$$

where the effective area $A = \pi a^2 \, f(V) = 3.7 \times 10^{-6}$ cm^2 for a = 11.3 μm and $f(V)$ = function of V number = 0.93, the effective length $L = (1 - e^{-\alpha\ell})/\alpha$ = 1.4×10^7 cm for α = 0.01 dB/km and ℓ = 175 km, and g is the gain coefficient assumed equal to fused silica values. The actual gain coefficients for fluoride glasses have not been determined.

The theoretical attenuation limits of an IR optical material for fibers are determined by the intrinsic attenuation processes of band-gap absorption, scattering and multiphonon absorption as described by the first, second, and third terms of Equation (5). A plot of these terms versus wavelength is a means by which the intrinsic limitations of an optical fiber can be visualized and determines the attenuation minimum. Such plots are often called "V" plots. An example of a V plot is shown in Figure 11.

Attenuation can also occur from the waveguide itself. Bending of a waveguide causes radiative losses due to the change in the angle of incidence of the guided energy with the core/clad interface, where some energy exceeds the critical angle for total internal reflection. Bending loss exponentially increases with decreasing radius of curvature. Other losses are caused by micro bending or waveguide irregularities that scatter the energy out of the waveguide.

Figure 12 is a diagram of many of the material and waveguide attenuation mechanisms.

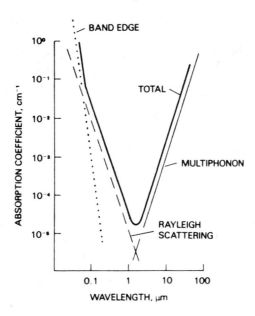

Figure 11. V-Plot (Reference 7)

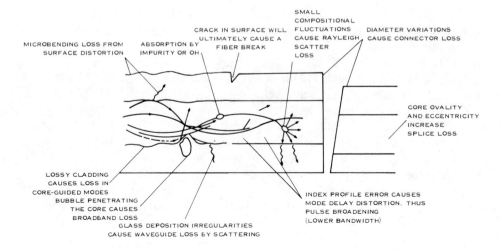

Figure 12. Optical Fiber Attenuation Mechanism (Reference 22)

1.3.2.2 Hollow Fibers

Attenuation in hollow waveguides results from the interaction of the electromagnetic energy with the walls of the waveguide causing resistive heating. The loss per unit length in a straight hollow metallic cylindrical waveguide is caused by the conductivity of the wall, σ, that can be related to the complex refractive index,

$$n = n_r - ik$$

where

$$k = \sigma\mu/2n_r\omega\epsilon_o$$
$$\omega = \text{frequency}$$
$$\alpha = \frac{2\omega k}{c} = \frac{\sigma\mu}{n_r c\epsilon_o}$$

and is given by:[10]

$$\alpha_{mn}(TM_{modes}) = \frac{\left(\frac{\omega\epsilon_o}{2\sigma}\right)^{1/2}}{a}$$

$$\alpha_{mn}(TE_{modes}) = \frac{\left(\frac{\omega\epsilon_o}{2\sigma}\right)^{1/2}\lambda^2}{a^3}\left(\frac{c_{mn}+m^2}{4\pi^2}\right)$$

(13)

where

$\quad$ a = waveguide radius (a > λ)

$\quad \epsilon_o$ = electric permittivity

ω and λ = frequency and wavelength

$\quad c_{mn}$ = mode geometric factor [nth root of $J_{m-1}(c_{mn}) = 0$].

When a cylindrical hollow metallic waveguide is bent, the loss in transmission scales as $1/R^2$ (R = radius of curvature) and is more severe than solid dielectric fibers and hollow metallic rectangular waveguides. The losses can be reduced by dielectric layers applied to the internal diameter of the metal waveguide because of the reduced surface impedance and admittance. The dielectric/metal waveguide losses are reduced and scale as $1/R$.[23]

The loss per unit length in a straight or bent hollow metallic rectangular waveguide is given by:[24]

$$\alpha_{mn} = \frac{m^2\lambda^2}{4a^3} R_{TE} + \frac{n^2\lambda^2}{4b^3} R_{TM} \qquad (14)$$

where

$\quad$ a = waveguide height

$\quad$ b = waveguide width

$\quad R_{TE}, R_{TM}$ = slope of reflectivity as a function of angle of incidence.

Hollow rectangular waveguides can also reduce their bending losses just as the circular waveguide by the use of dielectrics. In general, however, the bending loss of any hollow waveguide will exceed that of a solid dielectric fiber.

If the bend radii of a hollow rectangular waveguide is less than $R_c = 4a^3/\lambda^2$, then the electromagnetic energy follows "whispering gallery modes," which reflect off the upper wall and not the lower wall as it bends away from the light. The loss of the waveguide becomes independent of bend radii (less than R_c), mode number, length, and height and is given by:[24]

$$\alpha = \frac{A}{2} \theta \qquad (15)$$

where

$\quad$ A = absorptance at normal incidence of the waveguide wall

$\quad \theta$ = bend angle.

A twist can also be introduced into a hollow waveguide, but the loss increases as the square of the rate of twist.

The transmittance of a cylindrical dielectric hollow waveguide can also be given by:[25]

$$T = [R(\theta)]^{1/p} \quad (16)$$

where $R(\theta)$ = reflection coefficient as a function of angle =

$$\frac{1}{2}\left[\frac{\frac{\epsilon}{\epsilon_o}\cos\theta - \left(\frac{\epsilon}{\epsilon_o} - \sin^2\theta\right)^{1/2}}{\frac{\epsilon}{\epsilon_o}\cos\theta + \left(\frac{\epsilon}{\epsilon_o} - \sin^2\theta\right)^{1/2}}\right]^2 + \frac{1}{2}\left[\frac{\cos\theta - \left(\frac{\epsilon}{\epsilon_o} - \sin^2\theta\right)^{1/2}}{\cos\theta + \left(\frac{\epsilon}{\epsilon_o} - \sin^2\theta\right)^{1/2}}\right]^2 \quad (17)$$

where

ϵ = complex dielectric constant of wall

ϵ_o = dielectric constant of medium in waveguide, 1.0 if air

p = period of reflection in the fiber, or the inverse of the number of reflections per unit length, $\frac{1}{a\tan\theta}$.

1.3.3 Dispersion

Dielectric IR optical fibers have the potential for telecommunication applications. Their capacity to carry information depends on their delay distortion of the injected optical pulses. The delay distortion is the result of intermodal dispersion and intramodal dispersion, as shown in Figures 13 and 14. Intermodal dispersion results from the different path lengths and therefore different arrival times of two different modes excited by the same input pulse. The maximum time delay between two modes excited by the same input pulse is given by:[9]

$$\Delta T = \frac{Ln_1}{c}\left(\frac{n_1 - n_2}{n_2}\right) \quad (18)$$

where

L = length of fiber

c = vacuum speed of light.

Intramodal dispersion results from the dependence of a single mode's propagation constant on wavelength. Since the index of the material varies with wavelength, the mode group velocity also varies because of the spectral width of the source. Intramodal dispersion is given by:[9]

$$D = \frac{\lambda \Delta\lambda L}{c}\left(\frac{d^2n_1}{d\lambda^2}\right) (\text{ps/nm} \cdot \text{km}) \quad (19)$$

where

λ = peak wavelength

$\Delta\lambda$ = spectral width of source

$d^2n/d\lambda^2$ = material dispersion.

Intramodal dispersion is also called material dispersion, M, since $D = M \cdot \Delta\lambda$. The determination of the material dispersion zero, where $d^2n/d\lambda^2 = 0$, is very important for candidate materials for optical fibers, since it is the wavelength of maximum bandwidth for a multimode fiber. The material dispersion can be obtained using a Sellmeier formalism developed by Wemple.[26]

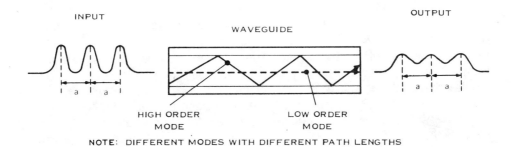

Figure 13. Intramodal Dispersion, Multimode Fibers Only

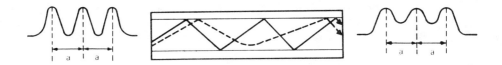

Figure 14. Intramodal Dispersion, Multimode and Single-Mode Fibers

The Sellmeier dispersion relation is given by the following:

$$n^2 - 1 = \frac{E_d E_o}{(E_o^2 - E^2)} - \frac{E_l^2}{E^2} \tag{20}$$

where

E_d = electronic oscillator strength
E_l = lattice oscillator strength
E_o = average excitation energy
E = energy in eV.

The first term in Equation (20) is the electronic contribution to the index and the second term is the lattice contribution. It is the negative lattice term that causes the material dispersion zero. The average excitation energy, E_o, is similar to average band-gap energy and scales as the optical band-gap, E_g^{opt}. The electronic oscillator strength, E_d, and the lattice oscillator strength, E_l, can be found from first principles related to the structure and chemistry of the material. The oscillator strengths E_l and E_d are not related to any particular resonance in the material but are averages of the lattice resonances and the electronic resonances, respectively. The single asymptotic term for the lattice contribution in Equation (20) is valid for frequencies much higher than those of the fundamental phonon frequencies. To obtain index values from the Sellmeier at lower frequencies, more than one lattice term would be required.

The values of E_d, E_l and E_o are found by manipulating Equation (20) into two linear relationships. For the lattice contribution, $E \ll E_o$, Equation (20) can be written as

$$n^2 - 1 = \frac{E_d}{E_o} - \frac{E_l^2}{E^2}. \tag{21}$$

Plotting $n^2 - 1$ versus $1/E^2$ gives a straight line where the slope equals $-E_o$ and the intercept equals E_d/E_o. For the electronic contribution, Equation (20) can be written as

$$\left[(n^2 - 1) + \frac{E_l^2}{E^2}\right]^{-1} = \frac{E_o}{E_d} - \frac{E^2}{E_d E_o}. \tag{22}$$

Plotting $[(n^2 - 1) + E_l^2/E^2]^{-1}$ versus E^2 gives a straight line where the slope equals $-1/E_d E_o$ and the intercept equals E_o/E_d. The intercepts of these two plots are reciprocals of each other.

Finally, the zero-dispersion wavelength is:

$$\lambda_o \simeq 1.63 \left(\frac{E_d}{E_o^3 E_l^2}\right)^{1/4} \mu m \tag{23}$$

and the material dispersion is:

$$M \simeq 1.54 \times 10^4 \frac{(E_d/E_0^3)}{(n\lambda^3)} - 2.17 \times 10^3 \, E_l^2 \, \lambda/n. \tag{24}$$

Ideally, the material dispersion zero should occur at a wavelength near the attenuation minimum. This results in a low-loss, high-bandwidth optical fiber link.

1.4 REFERENCES

1. T. Miya, Y. Terunuma, T. Hosaka, and T. Miyashita, "Ultimate low-loss single-mode fibre at 1.55 μm," *Electron. Lett.* **15** (February 1979), pp. 106–108.

2. L.G. DeShazer and C. Kao, eds., *Infrared Fibers (0.8 to 12 μm)*, *Proc. SPIE* **266** (1981).

3. L.G. DeShazer, "Advances in Infrared Fibers II," *Proc. SPIE* **320** (1982).

4. P. Klocek, ed., "Infrared Optical Materials and Fibers III," *Proc. SPIE* **484** (1984).

5. P. Klocek, ed., "Infrared Optical Materials and Fibers IV," *Proc. SPIE* **618** (1984).

6. P. Klocek, ed., "Infrared Optical Materials and Fibers V," *Proc. SPIE* **843** (1987).

7. M.G. Drexhage, K.P. Quinlan, C.T. Moynihan, and M. Saleh Boulos, "Fluoride Glasses for Visible to Mid-IR Guided-Wave Optics," *Advances in Ceramics, Vol. II: Physics of Fiber Optics*, B. Bendow and S.S. Mitra, eds., Am. Ceramic Soc., Columbus, Ohio (1981).

8. M.G. Drexhage, B. Bendow, O.H. El-Bayoumi, R.N. Brown, P.K. Banerjee, T. Loretz, C.T. Moynihan, J.J. Shaffer, P.A. Temple, and H.E. Bennett, "Progress in the development of multispectral glasses based on the fluorides of heavy metals," *Proceedings of the 13th Boulder Laser Damage Symposium*, Boulder, Colorado (1982).

9. S.E. Miller and A.G. Chynoweth, eds., *Optical Fiber Telecommunications*, Academic Press, Inc. (New York, 1979).

10. E. Garmire, T. McMahon, and M. Bass, *Applied Optics* **15**:1 (January 1976), pp. 145–150.

11. E.A.J. Marcatili and R.A. Schmeltzer, *Bell Sys. Tech. J.* (July 1964), pp. 1783–1809.

12. J.D. Jackson, ed., *Classical Electrodynamics*, John Wiley & Sons, Inc. (New York, 1975).

13. P. Klocek, *Materials Research Bulletin* **XI**:3 (1986), pp. 41–44.

14. J.I. Pankove, *Optical Processes in Semiconductors,* Dover Publications Inc., (New York, 1971).

15. R.H. Bube, *Electronics in Solids,* Academic Press (New York, 1981).

16. J.R. Gannon, "Optical fiber materials for operating wavelengths longer than 2 μm," *J. Non-Cryst. Solids* **42** (1980), pp. 239–246.

17. M. Sparks, "Theoretical Studies of Materials for High-Power Infrared Coatings," Xonics, Inc., Sixth Technical Report, Contract DAHC15–73–C–0127 (31 December 1975).

18. M. Sparks, "Continued Studies of Low-Loss Optical Fibers—Final Report," Scientific Research Center Report R3nr18354 (29 December 1983).

19. H. Vora, M. Flannery, and M. Sparks, "Tabulation of Impurity Absorption Spectra— Ultraviolet and Visible, Volume I," Xonics, Inc., Ninth Technical Report, Contract DAHC15-73-C-0127 (30 June 1977).

20. D.A. Pinnow, T.C. Rich, F.W. Ostermayer, Jr., M. DiDomenico, Jr., *Applied Physics Letters* **22**:10 (1973), pp. 527–529.

21. P. Klocek and M. Sparks, *Optical Engineering* **24** (1985), pp. 1098–1101.

22. From "The Western Electric Engineer."

23. M. Miyagi, K. Harada, Y. Aizawa, and S. Kawakami, *SPIE Proceedings* **484** (1984), pp 117–123.

24. E. Garmire, *SPIE Proceedings* **320** (1982), pp. 70–78.

25. T. Hidaka, T. Morikawa, and J. Shimada, *Journal of Applied Physics* **52**:7 (July 1981), pp. 4467–4471.

26. S.H. Wemple, *Applied Optics* **18**:1 (1979), pp. 31–35.

Section 2
IR Materials and Fibers

2.1 HALIDE GLASSES

A variety of materials, both crystalline and glassy, possess appropriate intrinsic properties (including minimum loss, material dispersion, and tensile strength) for potential application as mid-IR fibers. However, the major emphasis to date has focused on the fluoride glasses. A number of excellent reviews of recent work on fluorides can be found in the literature, including papers by Miyashita and Manabe,[1] Tran et al.,[2] and Drexhage.[3] In this section, a review is presented of the current status of fluoride glass technology, including the preparation and characterization of both bulk glasses and fibers. Material is presented that outlines the primary glass compositions, structure, preparation methods, and optical and physical properties for both bulk glasses and fibers.

2.1.1 Composition

Historically, the first fluoride glass synthesized, based on BeF_2, was discovered over 50 years ago. The fluoroberyllate glasses are of potential interest for high-energy laser applications because of their low linear and nonlinear refractive indices and low dispersion, but are unsuitable for practical handling and use because of their toxicity and hygroscopicity. For ultra-low-loss fiber applications, the fluoroberyllates show little advantage, since their IR absorption edge is only slightly shifted to a longer wavelength relative to oxide glasses. AlF_3-based glasses, synthesized by Sun in 1949,[4] transmit further in the IR, but are prone to devitrification and thus cannot be prepared into large bulk glass specimens or drawn into crystal-free fibers. The heavy metal fluoride (HMF) glasses, discovered by Lucas and Poulain in 1974, exhibit less tendency toward crystallization and higher IR transparency than the AlF_3-based glasses. The HMF glasses can be divided into two distinct categories. The first category involves fluorozirconate or fluorohafnate glasses, which contain ZrF_4 (or HfF_4) as the glass network former (50–70 mol %), BaF_2 as the primary network modifier (around 30 mol %), and one or more additional metal fluorides of the rare earths, alkalis, or actinides as glass stabilizers. Fluorozirconate glasses that are most resistant to devitrification always include four or more fluoride components. Typical ZrF_4-based glass compositions used for fiber fabrication are listed in Table 6.

TABLE 6. TYPICAL HEAVY METAL FLUORIDE GLASSES USED IN THE FABRICATION OF IR FIBERS (REFERENCE 2)

Core Composition (mole %)	Clad Composition (mole %)
$61ZrF_4$–$32BaF_2$–$3.9GdF_3$–$3.1AlF_3$	$59.5ZrF_4$–$31.2BaF_2$–$3.8GdF_3$–$5.5AlF_3$
$51ZrF_4$–$16BaF_2$–$5LaF_3$–$3AlF_3$–$20LiF$–$5PbF_2$	$53ZrF_4$–$19BaF_2$–$5LaF_3$–$3AlF_3$–$20LiF$
$53ZrF_4$–$19BaF_2$–$5LaF_3$–$3AlF_3$–$20LiF$ (+ –Cl, –Br, –I dopants)	$53ZrF_4$–$19BaF_2$–$5LaF_3$–$3AlF_3$–$20LiF$
$56ZrF_4$–$30BaF_2$–$5LaF_3$–$4ThF_4$–$5AlF_3$	$55ZrF_4$–$31BaF_2$–$5LaF_3$–$4NaF$–$5AlF_3$
$60ZrF_4$–$19BaF_2$–$6LaF_3$–$15NaF$	$57ZrF_4$–$12BaF_2$–$6LaF_3$–$25NaF$
$27ZrF_4$–$27HfF_4$–$23BaF_2$–$8ThF_3$–$4LaF_3$–$2AlF_3$–$3LiF$–$3NaF$–$3PbF_2$	$26ZrF_4$–$26HfF_4$–$23BaF_2$–$8ThF_4$–$4LaF_3$–$4AlF_3$–$4LiF$–$4NaF$–$1PbF_2$
$53ZrF_4$–$20BaF_2$–$4LaF_3$–$3AlF_3$–$20NaF$	Unclad
$43AlF_3$–$20BaF_2$–$20CaF_2$–$17YF_3$	Unclad

The second category of HMF glasses excludes ZrF_4 and HfF_4. Examples of this group include the AlF_3-based glasses[5] derived from Sun's original composition; the transition metal fluoride and rare-earth fluoride-based glasses, which offer interesting magnetic and lasing properties,[6,7] and the BaF_2/ThF_4-based glasses, which are transparent to around 7.0 μm as compared with 5.5 μm for the fluorozirconates.[8] Drexhage[3] has provided a summary of glass forming systems not containing ZrF_4 or HfF_4, as shown in Table 7.

Since most of these compositions are long and sometimes tedious to write repeatedly, acronyms are typically employed to describe the fluoride compositions. Table 8 lists some acronyms commonly used in the literature.

Selection of suitable glass compositions depends both on the projection of intrinsic optical properties and on important parameters relevant to fiber drawing, such as viscosity versus temperature curves and thermal analysis data. Fluoride glasses in particular are often unstable against crystallization. Differential scanning calorimetry can be used during both cooling and heating of the glass to determine both the glass transition temperature (T_g) and the temperature at which crystallization might occur (T_x) (Figure 15). By maximizing the value of $T_x - T_g$, it is possible to identify some candidate fluoride glass compositions for fiber drawing, as shown in Figure 16.

TABLE 7. GLASS-FORMING SYSTEMS NOT CONTAINING ZrF_4 OR HfF_4 AND GUIDE TO AVAILABLE DATA (REFERENCE 3)

System	Glass Formation Diagram	T_g, T_x	Density	Refractive Index	Optical Properties	Reference(s)
PbF_2–AlF_3	×	×	×	×	×	Kanamori et al., 1980; Shibata et al., 1980; Takahashi et al., 1981
PbF_2–MF_2–XF_3 (M = Mn, Co, Ni, Cu, Zn; X = Fe, V, Cr, Ga)	×	×	×	×	×	Miranday et al., 1979, 1980
AF–MF_2–XF_3 (A = Li, Na, K, Ag; M = Sr, Ca, Cd, Pb; X = Cr, Fe, V, Ga)	×	×				Miranday et al., 1979, 1980
BaF_2–CaF_2–AlF_3	×	×		×		Videau et al., 1979
BaF_2–ScF_3–MF_n (MF_n = NaF, YF_3, ThF_4)	×	×	×	×	×	Poulain et al., 1982
BaF_2–MnF_2–LnF_3 (Ln = Dy through Lu + Y)	×	×	×			LePage et al., 1982
BaF_2–ZnF_2–CdF_2	×	×	×		×	Matecki et al., 1982b
BaF_2–ZnF_2–LnF_3 (Ln = La through Lu + Y)	×	×	×		×	Fonteneau et al., 1980b
BaF_2–ThF_4–MF_2 (M = Mn, Zn)	×	×	×		×	Fonteneau et al., 1980a
BaF_2–ThF_4–LnF_3 (Ln = Tm, Yb Y)	×	×		×	×	Lucas et al., 1981
BaF_2–YF_3–AlF_3–CaF_2	×	×	×	×	×	Kanamori et al., 1981
BaF_2–YF_3–AlF_3–ThF_4	×	×	×	×	×	Poulain et al., 1981
MF_2–ZnF_2–AlF_3–ThF_4 (M = Mg, Ca, Sr, Ba)	×	×	×	×	×	Matecki et al., 1981
BaF_2–ZnF_2–MF_3–ThF_4 (M = Yb, Lu)	×	×	×	×	×	Slim, 1981; Drexhage et al., 1982a

TABLE 8. COMPOSITIONS OF HEAVY-METAL FLUORIDE GLASSES AND THEIR ACRONYMS AS USED IN THE TEXT

Acronym	Composition (mole %)
ZBT	$58ZrF_4$–$33BaF_2$–$9ThF_4$
HBT	$58HfF_4$–$33BaF_2$–$9ThF_4$
ZBL	$62ZrF_4$–$33BaF_2$–$5LaF_3$
HBL	$62HfF_4$–$33BaF_2$–$5LaF_3$
ZBGA	$60ZrF_4$–$32BaF_2$–$4GdF_3$–$4AlF_3$
ZBLA	$57ZrF_4$–$32BaF_2$–$3LaF_3$–$4AlF_3$
HBLA	$57HfF_4$–$36BaF_2$–$3LaF_3$–$4AlF_3$
ZBLAN	$55.8ZrF_4$–$14.4BaF_2$–$5.8LaF_3$–$3.8AlF_3$–$20.2NaF$
BZnYbT	$19BaF_2$–$27ZnF_2$–$27YbF_3$–$27ThF_4$
BZnLuT	$19BaF_2$–$27ZnF_2$–$27LuF_3$–$27ThF_4$
BZnYAT	$20BaF_2$–$29ZnF_2$–$14.4YF_3$–$14.4AlF_3$–$22.2ThF_4$
BYAT	$20BaF_2$–$29YF_3$–$29AlF_3$–$22ThF_4$

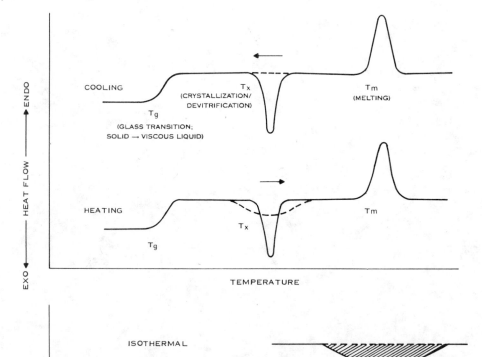

Figure 15. Thermal Analysis

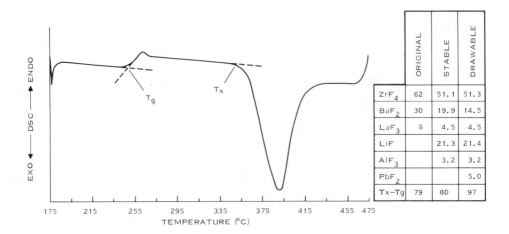

Figure 16. Fluoride Glass Fiber Materials, Thermal Analysis

2.1.2 Material Preparation

While BeF_2 easily forms a glass upon cooling from the melt, multicomponent fluoride glasses tend to have low viscosities at their liquidus temperature and tend toward crystallization. Moreover, fluoride melts are reactive with the atmosphere and with certain crucible materials, further increasing their susceptibility to crystallization and/or contamination. For these reasons, preparation of HMF glasses must be carried out under highly controlled conditions. Namely, nonreactive crucible materials, such as vitreous carbon, platinum, or gold, are required. Stringent atmospheric control during melting and quenching is necessary to prevent oxide and hydroxide contamination, which lead to undesirable mid-IR absorption and, possibly, nucleation and crystallization. Either inert atmospheres (N_2, Ar, He) or reactive gases such as CCl_4, SF_6, HF, CF_4, and BF_3 have been used to remove water and various oxide impurities from the melt. In many instances, batching and weighing of starting materials, melting, and casting have been carried out inside glove boxes or other controlled atmosphere chambers. Figure 17 shows some of the typical extrinsic sources of attenuation found in fluoride glasses,[9] including transition metal absorption and a strong OH fundamental near 3 μm. Both transition metals and rare earths cause problems in the 2- to 5-μm range of fluoride glasses. Table 9 shows the purification requirements to achieve a 0.01-dB/km loss in ZBLA glass.

Reactive atmospheres can be employed to convert the valence state of some transition ions to the species less absorbing in the IR region. For example, Tran[10] recently reported conversion of Fe^{2+} to Fe^{3+} in ZBLAN glass by oxygen pretreatment (Figure 18). Reactive atmospheric melting has been employed to reduce the water content of fluoride glasses. France et al.[11] have successfully lowered the OH fundamental to the vicinity of 30 dB/km in fluoride glass (Figure 19). Note that Figure 19 also shows a potentially interesting window near 2.6 μm at which losses as low as 10^{-2} dB/km might be achieved.

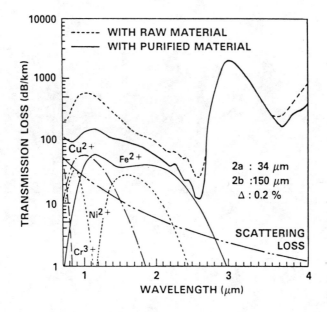

Figure 17. Spectral Attenuation of Fluoride Glass Fiber (NTT) (Reference 9)

TABLE 9. PURIFICATION REQUIREMENTS IN FLUORIDE GLASSES
IMPURITY LEVELS (PPB) CAUSING 0.01 dB/km LOSS

Impurity	Operational Wavelength		
	2.0 μm	3.0 μm	4.0 μm
Iron	0.11	5	25
Cobalt	0.08	10	50
Nickel	0.11	80	100
Copper	7.5	1000	>5000
Cerium	>5000	20	0.20
Praeseodymium	0.23	25	0.50
Samarium	10	1.7	1.43
Terbium	0.40	0.71	>5000

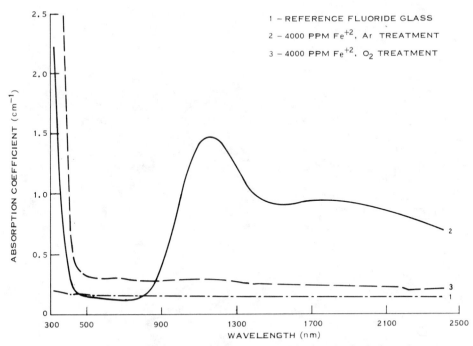

Figure 18. Effect of REDOX Melting Conditions on the Optical Absorption in Fluoride Glasses (Reference 10)

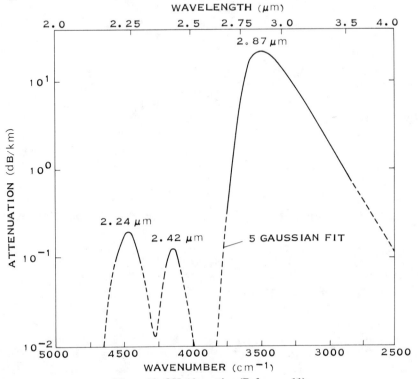

Figure 19. OH Absorption (Reference 11)

2.1.3 Physical Properties

2.1.3.1 *Optical Properties*

Perhaps the most important qualities of HMF glasses pertain to their extended transparency relative to silica in the 2- to 8-μm region. Figure 20 shows the characteristic extension of transmission in the IR of fluoride glass relative to silica without any sacrifice of optical quality in the visible or near-ultraviolet regions. The fundamental limitations in the ultraviolet transmission of glasses are the result of Rayleigh scattering caused by random fluctuations in density (and therefore index of refraction) frozen into the glass matrix upon cooling. Figure 21 shows some early work by Tran et al.[13] comparing light scattering in fluoride glasses to that of silica. The $1/\lambda^4$ dependence is that expected for Rayleigh behavior. More recently, Tran[10,14] reported light scattering levels in ZBLA + LiF (ZBLAL) glasses approaching theoretical values. These data are shown for some core and cladding glass compositions (Figure 22). Pb is used to raise the index of refraction of a fluoride glass so that it can be employed as a fiber core glass. Figure 23 shows the effect of a 3.7- and 5.0-percent addition of PbF_2 on the index of refraction of a ZBLAL glass over the spectral range of interest.

The IR edge is controlled by vibrational or multiphonon absorption and is not well quantified. Figure 24, taken from Drexhage,[3] shows the IR edge for various fluoride glass compositions, as well as superimposed on a fiber spectral attenuation curve. The scattering and IR edge data are then combined to determine the intrinsic projections of minimum loss for specific compositions. Typically, the heavy metal fluoride glasses are expected to possess minimum attenuation values of about two orders of magnitude below silica in the range of 10^{-2} to 10^{-3} dB/km.[16]

In addition to much lower optical losses potentially available in fluoride glasses, these materials show lower material dispersion over a broad spectral region than silica. Light pulses will not spread nearly as rapidly in time or spatial dimension in fluoride glasses. Miyashita and Manabe[1] have shown some typical data for BGZ and YABC fluoride glasses compared with silica (Figure 25). This lower dispersion characteristic of fluoride glasses has been confirmed by several investigators.

Table 10 is taken from a paper by Bendow.[17] It summarizes some of the other properties of interest for fluorides, citing current values versus projected intrinsic values.

One potentially interesting optical property of some fluoride glasses relates to their resistance to darkening under ionizing radiation. Figure 26 shows some of the NRL data obtained by Levin et al.[18] on irradiated ZBLA glass. The data show that bulk fluoride glasses exhibit excellent radiation hardness in the 2- to 5-μm range at doses as high as 10^7 rads from a ^{60}Co source. The response of optical fibers is not necessarily as good as these data since the optical pathlength is longer. However, even bulk optics that can sustain high radiation levels may be of interest. Note, however, that significant damage does occur in the ultraviolet and visible regions.

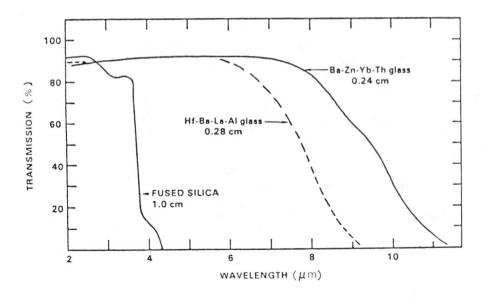

Figure 20. Comparison of Optical Transmission of Fluoride Glasses With Silica (Reference 12)

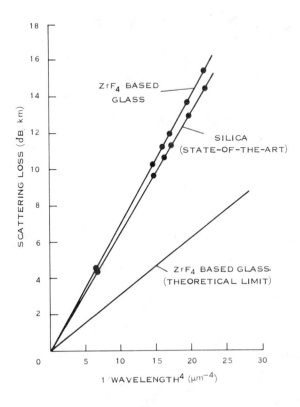

Figure 21. Scattering Loss Spectra for Bulk Fluoride Glasses (Reference 13)

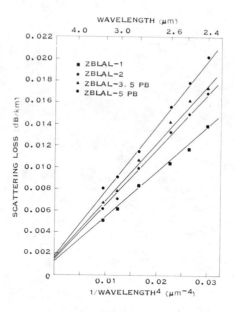

Figure 22. Light Scattering in Fluoride Glass and Silica (Reference 14)

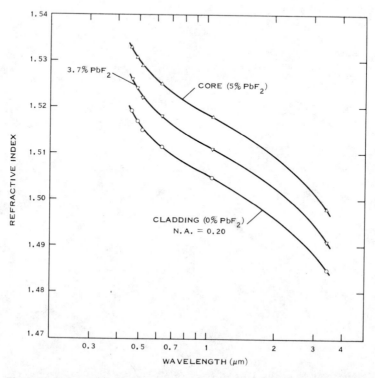

Figure 23. Refractive Index Spectra of Fluorozirconate Glasses (Reference 15)

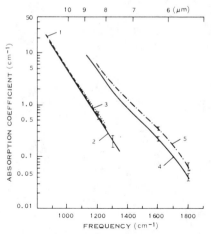
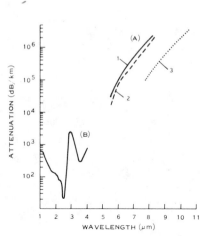
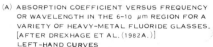

(A) ABSORPTION COEFFICIENT VERSUS FREQUENCY OR WAVELENGTH IN THE 6-10 μm REGION FOR A VARIETY OF HEAVY-METAL FLUORIDE GLASSES. [AFTER DREXHAGE ET AL. (1982A.)]
LEFT-HAND CURVES
1 $27\,YbF_3-27\,ThF_4-19\,BaF_2-27\,ZnF_2$
2 $27\,LuF_3-27\,ThF_4-19\,BaF_2-27\,ZnF_2$
3 $25\,LuF_3-25\,ThF_4-6\,GdF_3-19\,BaF_2-25\,ZnF_2$
RIGHT-HAND CURVES:
4 $57\,HfF_4-36\,BaF_2-3\,LaF_3-4\,AlF_3$
5 $57\,ZrF_4-36\,BaF_2-3\,LaF_3-4\,AlF_3$

(B) ABSORPTION COEFFICIENT IN DECIBELS PER KILOMETER VERSUS WAVELENGTH IN THE 1-10 μm REGION OBTAINED FROM MEASUREMENTS ON BULK GLASSES AND A ZrF_4-BaF_2-AlF_3-GdF_3 OPTICAL FIBER.
(A) BULK GLASSES: 1 FLUOROZIRCONATE
2 FLUOROHAFNATE
3 BaF_2/ThF_4 BASED (DREXHAGE ET AL., 1982A)
(B) FLUOROZIRCONATE GLASS FIBER (MITACHI AND MIYASHITA, 1982)

Figure 24. IR Edge for Various Fluoride Glass Compositions and Fiber (Reference 3)

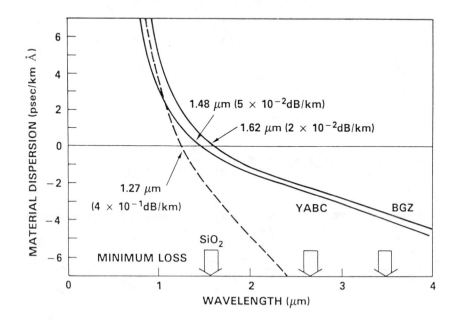

Figure 25. Material Dispersion of Fluoride Glasses (Reference 1)

TABLE 10. OPTICAL PROPERTIES OF FLUORIDE GLASSES (REFERENCE 17)

	Current	Projected
Refractive index	1.45–1.60	
Dispersion	$d^2/d\lambda^2 -5 \times 10^{-3} \mu m^2$ (IR)	
Absorption coefficient	2×10^{-5} cm^{-1} (HF) 10^{-4} cm^{-1} (6,348 Å)	2×10^{-8} cm^{-1} (HF & DF) 10^{-5} cm^{-1} (6,348 Å)
Rayleigh scattering	Similar to SiO_2	Lower than SiO_2
TIS	10^{-2} to 10^{-3} (DF)	$<10^{-3}$ (MID–IR)
Birefringence	$P_{11} - P_{12} \sim .005$	
dn/dT	(-0.9) to $(-1.5) \times 10^{-5}/°K$	
Thermal distortion	$1 \times 10^{-6}/°K$	$1 \times 10^{-7}/°K$
Laser damage (bulk)	Very good (DF)	Excellent
Surface finish	Very good	Excellent

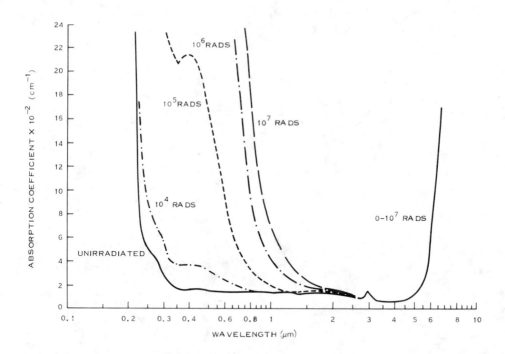

Figure 26. Absorption Spectra of Irradiated Fluorozirconate Glass (Reference 18)

2.1.3.2 Mechanical Properties

The hardness and fracture toughness of HMF glasses are generally lower than those of high silicates, but higher than those of chalcogenides, as shown in Table 11. Measured rupture strengths of bulk specimens (values up to 35 kpsi have been quoted in the literature) are, of course, a reflection of their surface condition rather than an indicator of the ultimate strength of the material. On the other hand, fracture toughness is an appropriate measure of intrinsic strength, and the values indicated in Table 11 imply strengths for HMF glasses that are roughly one-third of those of high silicates.

TABLE 11. MECHANICAL PROPERTIES OF FLUOROZIRCONATE GLASSES (REFERENCE 19)

Material	Hardness (kg/mm^2*)	Fracture Toughness (MPa·m$^{1/2}$)	Fracture Strength (MPa)	Thermal Expansion (10^{-7}/°C)
Fluorozirconate-type	225–250 (V)	0.25–0.27	20	150–180
Chalcogenide	100–250 (V)	0.20	20	120–250
Fused silica	~800 (V)	0.7–0.8	70	5.5
Calcium fluoride	~120–160 (Kp)	0.35	(40)	~180

*V = Vickers
Kp = Knoop

2.1.3.3 Other Selected Properties

Tables 12 and 13 present data on elastic, physical, and thermal properties of fluoride glasses.

TABLE 12. ELASTIC PROPERTIES OF FLUORIDES COMPARED WITH OXIDES (REFERENCE 19)

Material	Elastic Modulus MPa $\times 10^4$	(kpsi) $\times 10^3$	Bulk Modulus MPa $\times 10^4$	(kpsi) $\times 10^3$	Shear Modulus MPa $\times 10^4$	(kpsi) $\times 10^3$	Poisson's Ratio
Hf–Ba–La–Al fluoride glass	5.6	(8.1)	4.8	(6.9)	2.1	(3.1)	0.3
Zr–Ba–La–Al fluoride glass	5.5	(8.0)	4.8	(7.0)	2.1	(3.0)	0.3
Lead alkali silicate glass (50–60% PbO)	5.7	(8.3)	3.2	(4.6)	2.3	(3.4)	0.2
Fused silica	7.2	(10.5)	3.7	(5.3)	3.1	(4.5)	0.17

TABLE 13. PHYSICAL AND THERMAL PROPERTIES OF FLUORIDES (REFERENCE 19)

	Current	Projected
Fracture strength	10 kpsi	300 kpsi
Stress corrosion	N = 10 to 40	N ≥ 100
Hardness	300 to 400 (Vickers 50 g, 15 s)	>400
Softening temperature	300 to 400°C	>400°C
Crystallization temperature	400 to 500°C	>500°C
Thermal expansion	1.4 to 1.9 $\times 10^{-5}$/°C	<1.4 $\times 10^5$/°C
Thermal conductivity	Similar to silicates	—
Poisson's ratio	0.18 to 0.30	—
Thermal endurance (calculated)	ΔT–50–70°C	ΔT > 500°C

2.1.4 Waveguide Fabrication

In general, the same two basic approaches for fiber drawing used for oxide glasses can also be used for fluorides; namely, crucible drawing from the melt[20,21] and drawing of a solid preform or node of material such that the geometry of the fiber matches that of the larger bulk cylinder. Each approach has its advantages and disadvantages. Crucible drawing is a potentially continuous process, whereas a preform provides a fixed amount of material. The index profile is more difficult to control with a crucible draw, and crystallization at the exit nozzle has been a problem. Most fluoride glasses drawn by crucible have been single component with a polymer cladding[22] such as shown schematically in Figure 27. Precise temperature control is needed here because of sharp variations in viscosity with temperature in the fluoride glass system.

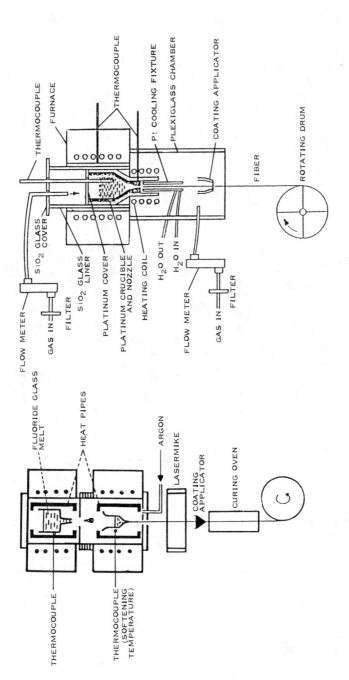

Figure 27. Fiber Drawing (Reference 20, left; Reference 23, right)

47

Most efforts have focused on the fabrication of preforms of a fluoride glass core and cladding, which is subsequently drawn into a fiber. Workers at Nippon Telegraph and Telephone (Mitachi et al.[24]) invented the built-in casting process in which the molten cladding glass is poured into a cooled mold and then inverted, leaving glass solidified on the walls of the mold. The core glass is then added as a second step to create a solid preform. This process is shown schematically in Figure 28. Tran et al.[26] at the Naval Research Laboratory improved the NTT process by going to a rotational casting process in which the metal mold is spun at several thousand RPM with the cladding glass in place and gradually cooled to produce a bubble-free, extremely uniform tube of fluoride glass. The core glass melt is then added from above or drawn from below to form the finished preform. The rotational casting process is shown schematically in Figure 29. Figure 30 is a cross sectional plot of the index of refraction measured at two points several inches apart on a rotationally cast preform. Note the uniformity of the core and cladding. In contrast, the built-in casting technique tends to suffer from tapering of the cladding tube.

Recently, a reactive vapor transport (RVT) technique has been reported[27] for fluoride glasses in which a ZBLAL substrate tube is doped internally with a heavier halide species (Figure 31). Cl, BR and I have been introduced into the inner wall to generate both graded index and single-mode structures. A typical index profile for an RVT multimode preform is shown in Figure 32. In principle, the RVT process could be reduced completely to a vapor phase technique by initial deposition of the cladding glass. The problem with CVD techniques in the fluoride glasses lies in the low vapor pressures available for many of the fundamental constituents. At present, no one has reported CVD deposition of heavy metal fluoride glasses by a process analogous to that used for SiO_2 fibers. The most promising fluoride glass system for the vapor phase approach appears to be BeF_2, as described in U.S. Patent #4,378,987 (Miller, et al., 1983), but the toxicity of this system is a drawback to its implementation.

The threat of crystallization of the preform during fiber drawing requires that the process be rapid. A narrow heat zone is needed to ensure that the preform is held at high temperature for the minimum period of time. Figure 33 shows a typical resistance-heated furnace for drawing fluoride fibers with a localized heat zone. Draw rates for fluoride glass fibers are similar to silica fibers and kilometer-continuous lengths of fluoride glass fibers have been achieved.

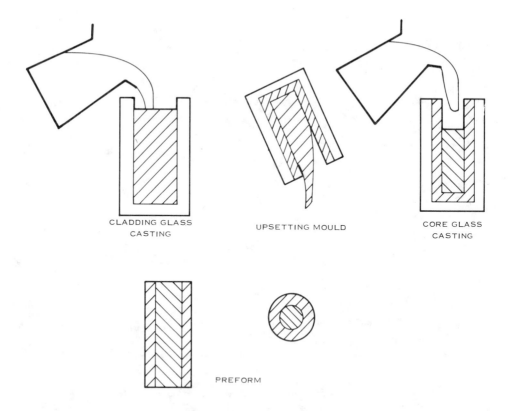

Figure 28. Adding Core Glass to Create Solid Preform (Reference 25)

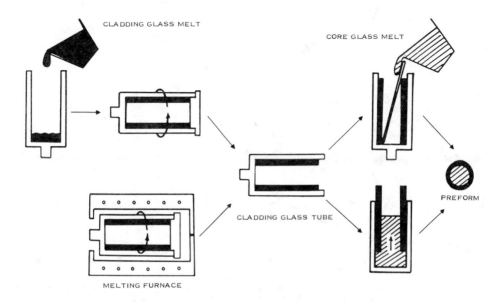

Figure 29. Rotational Casting Process (Reference 26)

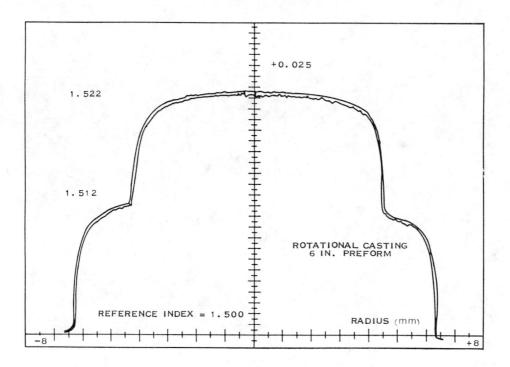

Figure 30. Index of Refraction Across Preform (Reference 2)

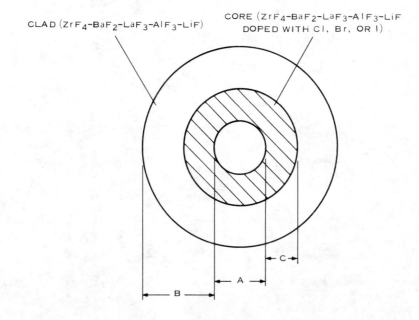

Figure 31. Reactive Vapor Transport Fluoride Glass Fiber (Reference 27)

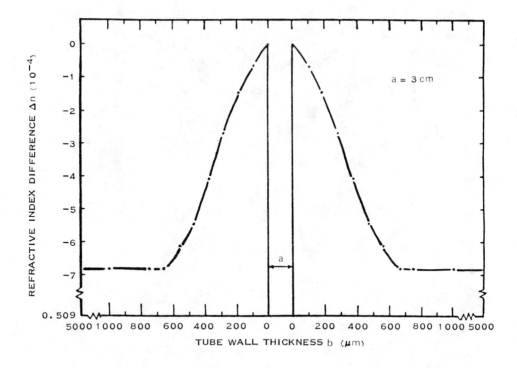

Figure 32. Refractive Index Versus Tube Wall Thickness for RVT Fibers (Reference 27)

Figure 33. Localized Heat Zone Furnace for Drawing Fluoride Glass Fibers

2.1.5 Waveguide Properties of Fluoride Fibers

The most critical parameter for fluoride fibers is their optical transparency. Although potentially more transparent than silica, losses still remain high because of extrinsic impurity absorption. Figure 34 shows data reported by Tran et al.[9,28] for two classes of HMFG fibers. The minimum loss near 2.6 μm is about 2 dB/km. The loss trend for fluorides is still decreasing, as shown in Figures 35 through 39. Silica, on the other hand, has reached its intrinsic attenuation limit of 0.16 dB/km. These multimode fibers were measured using straightforward ac techniques shown schematically in Figure 40. It is clear from analysis of the data that loss reductions in the future will depend on reduction of water in the glass, elimination of transition metals, and a further decrease of scattering in the fiber, especially at the core-cladding interface. Purification of starting materials is currently a key factor, as is OH reduction during processing and melting.

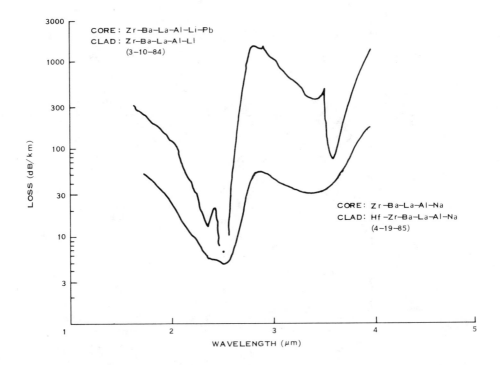

Figure 34. Attenuation of Fluoride Glass Fibers (Reference 28)

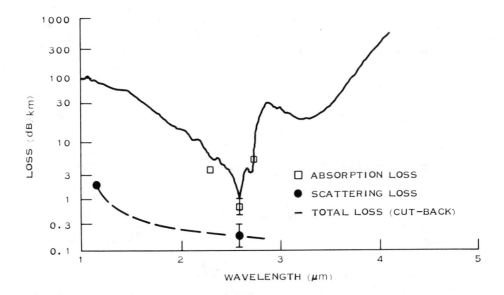

Figure 35. Loss Spectrum of a Fluoride Glass Fiber Produced at NRL (Reference 29)

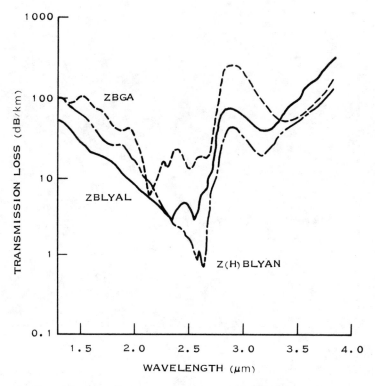

Figure 36. Transmission Loss Spectra for Step-Index Multimode Fibers (Reference 30)

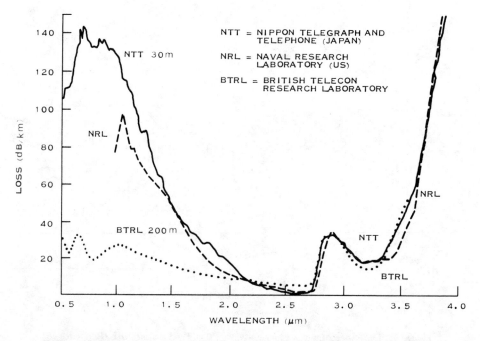

Figure 37. Comparison of Losses for Various Fluoride Glass Fibers (Reference 31)

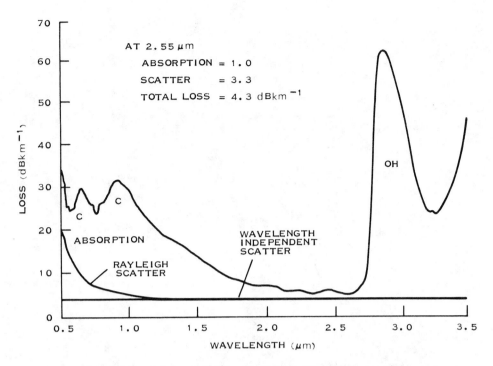

Figure 38. Breakdown of Total Loss in Fluoride Fiber (Reference 31)

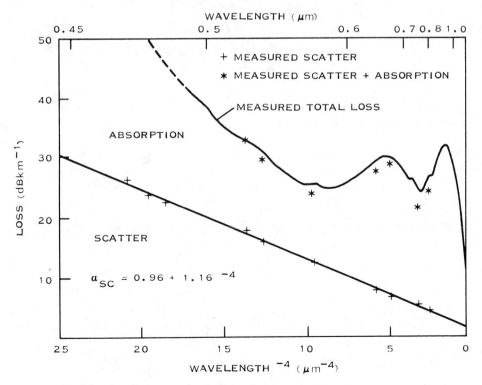

Figure 39. Measured Scatter Loss of Fluoride Fiber (Reference 31)

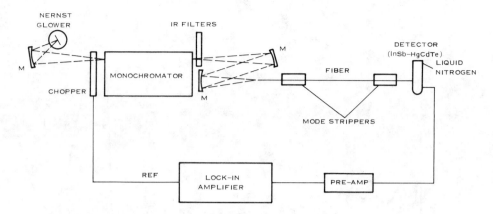

Figure 40. IR Fiber Spectral Loss Measurement System (Reference 15)

2.1.6 Mechanical Properties of Fluoride Fibers

The strength of HMF fibers has been investigated recently by several groups. British Telecom[32] reports breaking strengths around 70 kpsi (strains about 1 percent), as shown in Figure 41. The UCLA group reports strengths of up to 100 kpsi for freshly drawn, Teflon-coated fibers of 150 μm diameter. Fractographic analysis conducted at Sandia Laboratories[33] indicates relatively high values of fracture toughness for HMF fibers, namely $K_{IC} \sim 0.5$ MPa m$^{1/2}$. This value compares favorably with other glasses and leads to projected strengths for pristine HMF fibers in the 10^6 psi range. Assuming similar results for fluorides as for silicates, where the maximum strength attained in fibers is about one-third the theoretical limit, the practical strengths in carefully prepared HMF glass fibers can approach 500 kpsi.

Many studies have been conducted to assess the effect of humidity and/or water on HMF fiber strength. Figure 42 shows the rapid degradation of strength observed with ZBLA fibers immersed in water.[32] Humidity tests conducted at UCLA showed that strengths of Teflon-FEP-coated HMF fibers decrease significantly in wet atmospheres. In particular, water was found to permeate rapidly through FEP Teflon and lead to surface attack of the fibers. These conclusions are consistent with the susceptibility to corrosion and surface attack observed with bulk samples.

It is, therefore, evident that the theoretical strength of HMF fibers is extremely high. However, it is also evident that precisely controlled processing conditions and a means of hermetically sealing the fiber immediately after it is drawn will be required to realize the high potential strengths of HMF fibers.

Comparative studies in water and air indicate a decrease of 10 to 20 percent in the fracture toughness of fluorozirconates in water, which suggests that they

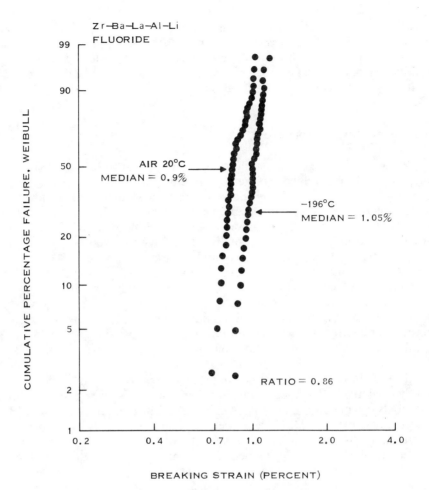

Figure 41. Strength of Fluoride Glass Fibers (Reference 32)

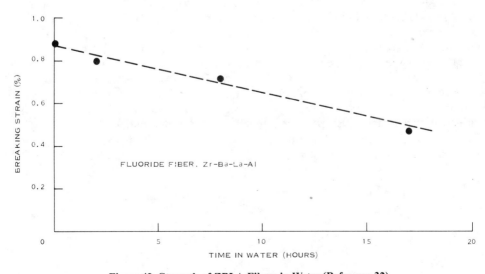

Figure 42. Strength of ZBLA Fibers in Water (Reference 32)

are susceptible to stress corrosion in aqueous environments. Delayed failure measurements on flexure bars in air indicate substantial variations of the stress corrosion coefficient n from sample to sample; some n values lie between 8 and 14, while others are greater than 50. Measured n values did not correlate systematically with OH content in the glass.

2.1.7 Durability and Toughness

While the surfaces of most fluoride glasses appear to be highly stable over long periods of time in laboratory environments, their potential susceptibility to attack in aqueous environments is a concern. For example, studies have shown that the solubility of typical fluorozirconates is many orders of magnitude lower than that of high silicates. In studies conducted by Simmons et al.,[34] for example, normalized leach rates (g/cm^2 · d) of 10^{-2} to 10^{-3} were obtained for ZrF_4-BaF_2-LaF_3-AlF_3 glass, compared with 10^{-7} to 10^{-8} for Pyrex (Figure 43). Figure 44 shows water absorption of various coated fluoride glasses. As studies have been limited to just a few compositions, it is not known whether others might possess lower solubilities. Moreover, the role of stress corrosion (i.e., crack growth under load in aqueous environments) in the durability of HMF glass is still uncertain. However, in contrast to their susceptibility to aqueous corrosion, HMF glasses appear to be highly resistant to fluorinating agents such as HF, F_2, and UF_6.

The potential susceptibility of fluoride glasses to aqueous attack, coupled with their relatively low fracture toughness and hardness, has spurred an interest in methods for toughening these glasses, some of which are indicated in Table 14. Diamond-like carbon (DLC) has been applied successfully to fluoride glass substrates. Preliminary hermeticity studies conducted at Catholic University indicate that the DLC coating reduces OH permeation to the surface of the glass. Once optimized, hermetic coatings such as DLC could lock in the pristine strength of the glass and protect glass surfaces indefinitely from corrosion.

Another approach to glass toughening involves surface treatments of various types. Thermally induced hardening of surfaces of Ba/Mn fluoride glasses has been reported, in which up to three exothermic transitions accompanied by cumulative increases in hardness up to 50 percent were observed. Ion exchange and compressive coatings are other well-known methods for toughening conventional glasses, but these have yet to be applied to HMF glasses. Formation of glass ceramics using heat and pressure is another approach that has not yet been systematically explored. Appropriate choices of composition can also yield glasses with substantially improved mechanical characteristics. And, although the detailed dependence of strength, hardness, and durability on composition, processing conditions, and surface preparation has yet to be established, current research suggests that variations in these parameters can be exploited to toughen HMF glasses.

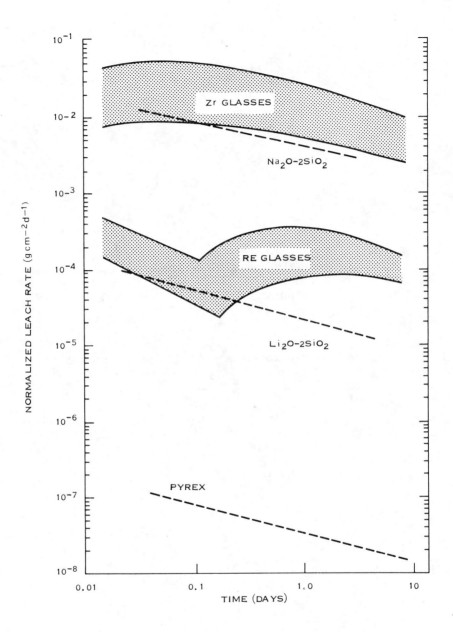

Figure 43. Leach Rates of ZBLA Glass With Pyrex (Reference 34)

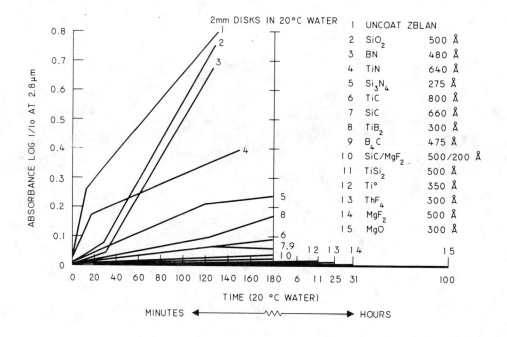

Figure 44. Water Absorption of Coated Fluoride Glasses (Reference 35)

TABLE 14. PROMISING METHODS FOR FURTHER TOUGHENING FLUORIDE GLASSES

- **Hermetic coatings**

    Diamond-like carbon (D)

    Metals (for optical fibers)

    Silicon, silicon nitride, boron nitride

- **Surface treatment**

    Thermally-induced recrystallization (D)

    Ion-exchange

    Compressive coatings (e.g., by CVD)

D = Demonstrated

2.2 CHALCOGENIDE GLASSES

2.2.1 Composition

Chalcogenide glasses are vitreous materials composed of the chalcogen elements of Group VI of the periodic table—in particular, sulfur, selenium, and tellurium. These elements are usually mixed with elements of Groups IV and V to form the familiar compound glasses. The chalcogenide glasses are semiconductor materials that tend to have a featureless band-gap owing to the amorphous structure of glasses. Elemental glasses of chalcogenides, such as Se, exist. Very stable and durable glasses can be found in a wide range of compositions from binaries to four or more components. These glasses appear to have network structures and to be predominantly covalently bonded. Table 15 lists the many chalcogenide compositions, not all of which have been investigated for IR fiber optics. Compositional diagrams of some of the glasses appear in Figures 45 through 48, where the circled regions represent the glass-forming compositions.

TABLE 15. CHALCOGENIDE GLASS COMPOSITIONS; VARIOUS ATOMIC RATIOS

As-S
As-Se
Ge-S
Ge-Se
Ge–As–S
Ge-Sb-Se
Ge–As–Se
Ge–As–Te
Ge–La–Ga–Se
Ge–P–S
Ge–P–Te
Ge–Se–Te
Ge–P–Se
Ga-Sb-Se-I
As–Se–Te

1	2	3
Si	P	S
Ge	As	Se
Sn	Sb	Te

Various atomic ratios of any permutation of one element from each column 1, 2, and 3

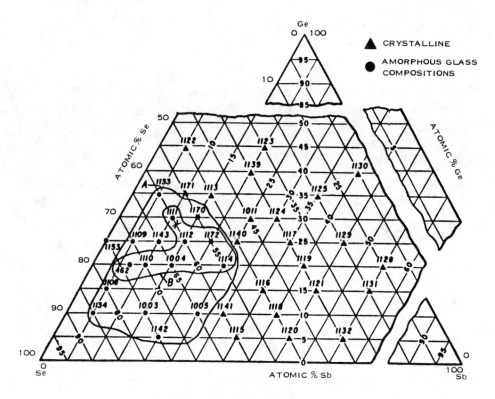

Figure 45. Glass-Forming Compositions of GeSbSe; Four-Digit Numbers Are Sample Identification (Reference 1)

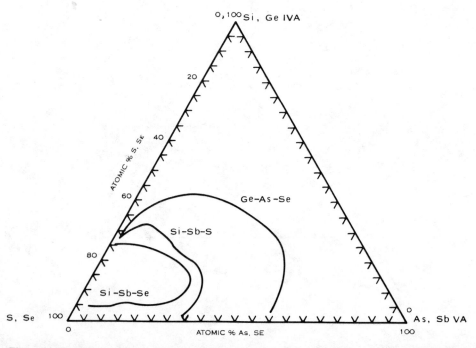

Figure 46. Glass-Forming Composition Region of Systems SiSbS, SiSbSe, and GeAsSe (Reference 2)

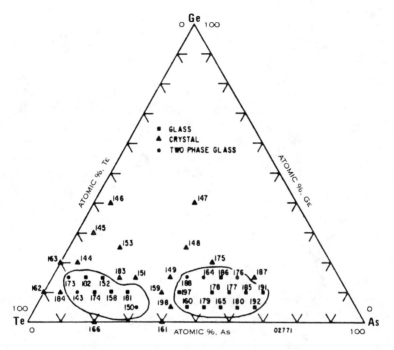

Figure 47. GeAsTe Composition Diagram; Numbers Are Sample Identification (Reference 3)

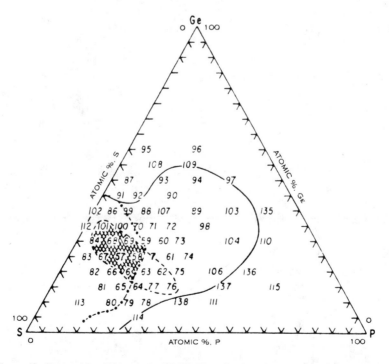

Figure 48. Composition Diagram for GePS Glass System, Shaded Area Indicates the Most Stable Glass-Forming Region; Numbers are Sample Identification (Reference 4)

2.2.2 Materials Preparation

Generally, chalcogenide glasses are formed by a bulk batch process from elemental raw materials. Usually, high-purity (99.999 percent or better) elements are weighed out in the appropriate amounts for the desired composition and placed in a quartz vessel. The quartz vessel is evacuated to a pressure of 10^{-4} torr or less and sealed. The vessel is then heated above the liquidus until a liquid mixture is formed. The mixture is agitated to homogenize the glass. The vessel is quenched and the bulk glass removed. Typical heating temperatures reach 1,000°C and cycles may run as long as tens of hours to achieve proper homogeneity. Figure 49 is a diagram of a typical quartz vessel and rocking furnace used in the fabrication of chalcogenide glasses.

Chalcogenide glasses are also formed by evaporation processes, but this has only recently been used in the fabrication of fibers. Although chemical vapor deposition (CVD) has been reported for some compositions, this development will be discussed in the Preform Fabrication section.

Optical quality chalcogenide glasses must be homogeneous to minimize light scattering. This homogeneity is accomplished by the thermal soak and mixing described above. Large defects in the glass, such as bubbles or particles, are avoided by the evacuation of the quartz vessel and the vapor transport of the raw materials through a frit inside the vessel. Metallic impurities are minimized by use of high-purity elemental raw materials, but, for low-loss optical fibers, improvement over 99.999 percent will be necessary. Anionic molecular impurities are the current limiting factor in achieving low-loss fibers. Oxygen, hydrogen, and water are the primary impurities in chalcogenide glasses. The IR transparency of the glass is limited by the absorption bands created by the vibrational characteristics of the impurities in the glass. The removal of hydrogen and oxygen from the glass is a primary development effort. The presence of these impurities dissolved in the glass is from the fabrication technique; thus, the development efforts are in CVD processes and purification methods. The sources of these anion impurities are dissolved species in the raw materials, absorbed species on the surfaces of the raw materials and vessel wall, and dissolution of the vessel wall. Figure 50 is a schematic of a distillation process that has had some success in the removal of oxygen from chalcogenide glass. A successful alternative approach to removing oxygen is the addition of small amounts of aluminum to the glass melt.[5] Since Al_2O_3 is a thermodynamically preferred oxide, the oxygen in the glass bonds to the Al forming Al_2O_3, whose IR absorption bands occur at wavelengths outside the transparency region of the glass. Some success has also been achieved in the removal of hydrogen impurity, which usually bonds to the chalcogen (i.e., H_2Se) in the glass, by a reaction process. For Ge-P-S glass, hydrogen impurity levels have been reduced by flowing S_2Cl_2 over the glass melt.[7] The influence of impurities is shown in Subsection 2.2.3, Physical Properties.

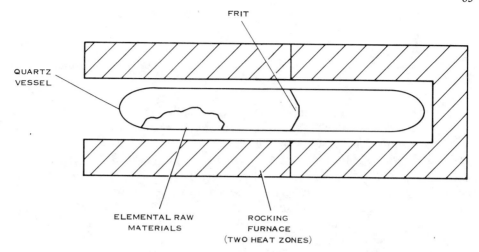

Figure 49. Fabrication of Chalcogenide Glasses

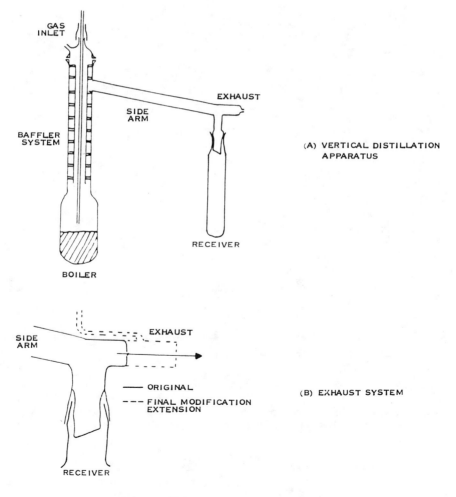

Figure 50. Distillation Process Schematic Diagram (Reference 6)

2.2.3 Physical Properties

The chalcogenide glasses offer a wide variety of compositions and, therefore, a wide variety of physical properties.

The IR spectra of many chalcogenide glasses are shown in Figures 51 through 58. The influence of impurities is labeled in each of the spectra. Absorption coefficient data for some of the glasses appear in Table 16. Much more detailed absorption data for fibers are provided in Subsection 2.2.5, as are material dispersion data. Table 17 lists many of the optical, thermal, and mechanical properties of the chalcogenide glasses used for IR optical fibers. Chalcogenide glasses are reasonably stable against devitrification and, therefore, can be easily processed into fibers with minimal extrinsic light scattering. The chemical durability of chalcogenide glasses is such that they are resistant to attack in water and acids but show appreciable dissolution in bases.

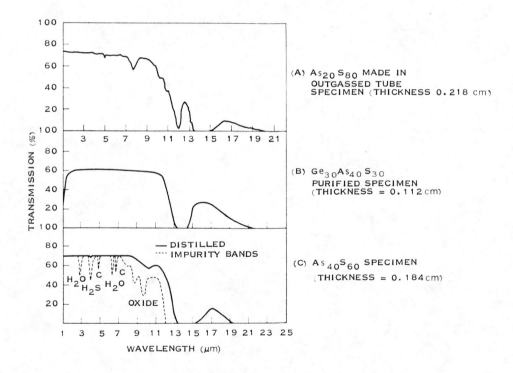

Figure 51. IR Transmissions of Purified Sulfide Glasses in Atomics Percent (Dashed Curves Show Impurity Bands) (Reference 8)

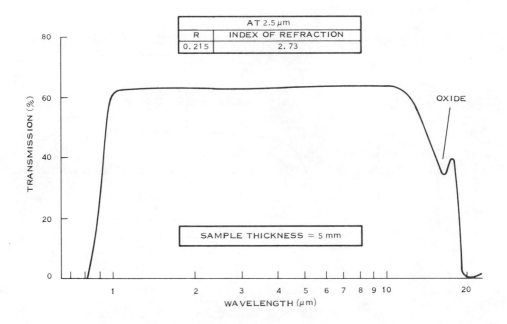

Figure 52. IR Spectra of As_2Se_3 (R = Reflection Coefficient) (Reference 9)

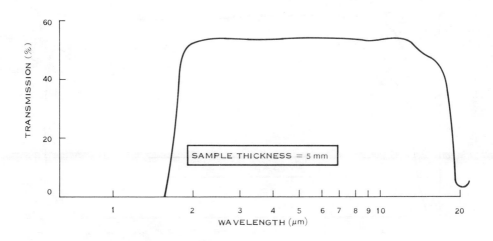

Figure 53. IR Spectra of As_2SeTe_2 (R = Reflection Coefficient) (Reference 9)

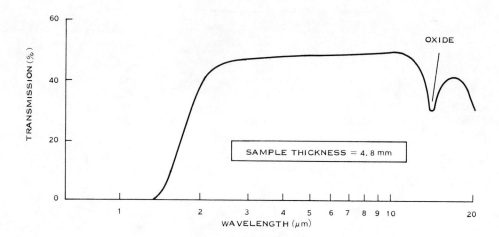

Figure 54. IR Spectra of $Ge_{10}As_{50}Te_{40}$ (R = Reflection Coefficient) (Reference 9)

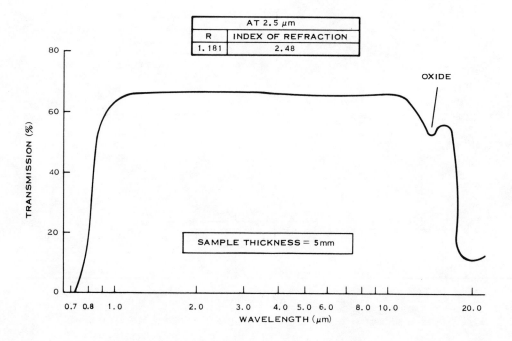

Figure 55. IR Spectra of $Ge_{15}As_{10}Se_{75}$ (R = Reflection Coefficient) (Reference 9)

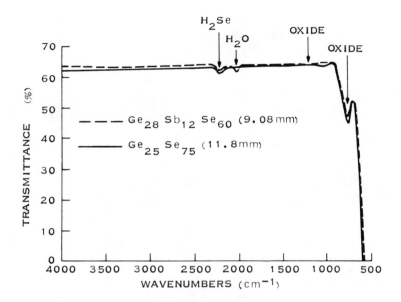

Figure 56. IR Spectra of $Ge_{28}As_{12}Se_{60}$ and $Ge_{25}Se_{75}$ (Reference 10)

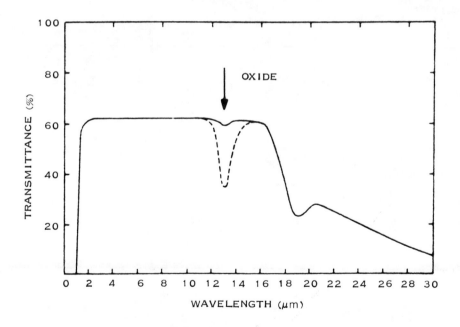

Figure 57. Optical Transmission of 2-mm-Thick Disk of GeSeTe Glass, Showing (Solid Curve) Removal of Absorption Band at 13.0 μm by Heating Glass in Hydrogen (Reference 11)

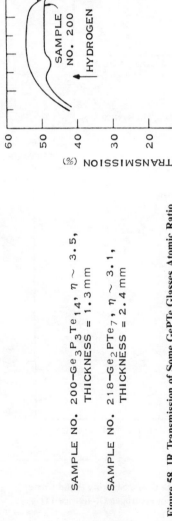

SAMPLE NO. 200—$Ge_3P_3Te_{14}$, $\eta \sim 3.5$, THICKNESS = 1.3 mm

SAMPLE NO. 218—Ge_2PTe_7, $\eta \sim 3.1$, THICKNESS = 2.4 mm

Figure 58. IR Transmission of Some GePTe Glasses Atomic Ratio (Reference 3)

TABLE 16. ABSORPTION COEFFICIENTS IN BULK CHALCOGENIDE GLASSES

Reference	Composition (Atomic percent)	Absorption Coefficient cm^{-1} (dB/m)				
		2.7 μm	3.39 μm	9.27 μm	10.6 μm	9.3 to 11.4 μm
9	$As_{40}S_{60}$	—	3×10^{-3} (1.3)	—	—	0.5 (217)
9	$As_{40}Se_{60}$	—	7×10^{-3} (3.0)	—	—	1.2×10^{-2} (5.2)
10	$Ge_{28}Sb_{12}Se_{60}$	7.0×10^{-4} (0.3)	—	3.0×10^{-3} (1.3)	1.9×10^{-2} (8.3)	—
9	$As_{40}Se_{40}Te_{20}$	—	1.2×10^{-2} (5.2)	—	—	2.4×10^{-2} (10.4)
9	$Ge_{15}As_{10}Se_{75}$	—	6×10^{-3} (2.6)	—	—	1×10^{-2} (4.3)
9	$Ge_{10}As_{50}Te_{40}$	—	0.11 (47.8)	—	—	6.6×10^{-2} (28.7)

TABLE 17. PHYSICAL PROPERTIES OF CHALCOGENIDE GLASSES

Composition (Atomic Percent)	Glass Transition Temperature (°C)	Thermal Expansion ($\times 10^{-6}$/°C)	Hardness (kg) (mm)^2	Young's Modulus (psi)	Poisson Ratio	Transmission Range (μm) 10% cutoff	$\frac{dn}{dT}$ ($\times 10^{-6}$ K^{-1})
Ge$_{25}$S$_{75}$	260 (Reference 12)	25 (Reference 12)	(Knoop) 130 (Reference 12)	—	—	0.5 to 11	—
Ge$_{25}$Se$_{75}$	240 (Reference 10)	~30 (Reference 10)	(Vickers) ~160 (Reference 10)	~3.1 × 10^6 (Reference 10)	~0.24 (Reference 10)	0.8 to 16	—
Ge$_{28}$Sb$_{12}$Se$_{60}$	300 (Reference 10)	14.1 (Reference 10)	(Vickers) 160 (Reference 10)	3.1 × 10^6 (Reference 10)	0.24 (Reference 10)	1.0 to 16	55 (Reference 10)
As$_{40}$S$_{60}$	182 (Reference 9)	24.0 (Reference 14)		2.3 × 10^6 (Reference 10)	0.31 (Reference 14)	0.6 to 11	−8.6 (Reference 14)
As$_{40}$Se$_{60}$	179 (Reference 9)	21.0 (Reference 14)	(Vickers) 156 (Reference 13)	2.9 × 10^6 (Reference 10)	0.29 (Reference 14)	0.8 to 16	223 (Reference 14)
As$_{40}$Se$_{20}$Te$_{40}$	114 (Reference 9)	21.5 (Reference 9)	(Knoop) 109 (Reference 13)	—	—	1.6 to 18	—
Ge$_{15}$As$_{10}$Se$_{75}$	150 (Reference 9)	27.1 (Reference 9)	(Knoop) 115 (Reference 13)	—	—	0.8 to 16	—
Ge$_{10}$As$_{50}$Te$_{40}$	170 (Reference 9)	12.7 (Reference 9)	(Knoop) 146 (Reference 13)	—	—	1.6 to 25	—
Ge$_{30}$As$_{15}$Se$_{55}$	351 (Reference 13)	12.8 (Reference 13)	244 (Reference 13)	~2.7 × 10^6 (Reference 13)	~0.27 (Reference 13)	0.8 to 16	—

2.2.4 Waveguide Fabrication

IR optical fibers of chalcogenide glasses, like silica-based fibers, can be fabricated easily because of the viscoelasticity naturally inherent in glass materials. Two general techniques have been used: one draws fibers directly from the molten glass, while the other draws fibers from solid glass preforms. Typical draw speeds are several meters per minute.

2.2.4.1 Preform Fabrication

An optical fiber preform is a solid precursor to the fiber that is heated and drawn, resulting in optical fibers with a geometry proportional to the preform. Chalcogenide glass preforms have been fabricated for a rod-in-tube type fiber drawing. Rods of the glass are formed in the quartz vessels where the raw materials were reacted into glass, that is, by selecting the dimensions of the reaction vessel so that the glass assumes a rod shape. IR fibers have been fabricated from these rods by heating and drawing, as shown in Figure 59. Rods of the glass have also been formed by extruding the glass through a die while heating the glass above its T_g. The fabrication of stepped-index fibers requires the glass rod to be placed inside a glass tube, which forms the clad. These glass tubes, formed by extrusion and rotational casting, are shown in Figure 60. Attention must be given to the surfaces of these preforms because the glass surface will dissociate and oxidize, causing absorption and scattering attenuation in the fibers. This is why inert atmosphere or vacuum conditions are maintained during preform fabrication. Preforms can also be chemically or mechanically polished to improve their surface quality. Both square and cylindrical fibers have been fabricated from these preform methods, as shown in Figure 61.

Chemical vapor deposition (CVD)[16,17,18] is in preliminary development for chalcogenide glass fiber-optic preforms. Generally, vapors of the halides (e.g., Cl, Br) or hydrides (H) of the desired elements (e.g., $GeCl_4$, S_2Cl_2, Se_2Cl_2) are drawn into a reactor tube of quartz or Teflon at flow rates determined to give the desired glass composition. The vapors can be decomposed thermally to get stoichiometric compounds (e.g., $GeSe_2$) or by a plasma process (RF or microwave) to get nonstoichiometric compounds (e.g., $Ge_{20}Se_{80}$). A plasma-enhanced process (PECVD) will, however, generally incorporate the other element present in the reactor (e.g., Cl or H). Considering most chalcogenide glasses of interest are nonstoichiometric, the impurity incorporation of PECVD is an important source of attenuation. Because oxygen, often the most harmful impurity in chalcogenide glasses for 8- to 12-μm transmission, is not present in the PECVD process, PECVD can offer a significant advantage.

The CVD process requires the glass to be deposited inside a tube. The cladding will be deposited first if desired and then the core glass inside the clad. The outside tube can then be removed resulting in the final preform for fiber drawing.

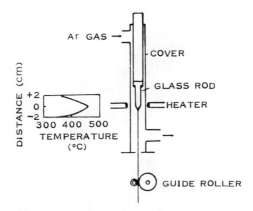

Figure 59. Apparatus for Fiber Drawing and Typical Data for Temperature Gradient at Neck-Down Region of Glass Rod (Reference 12)

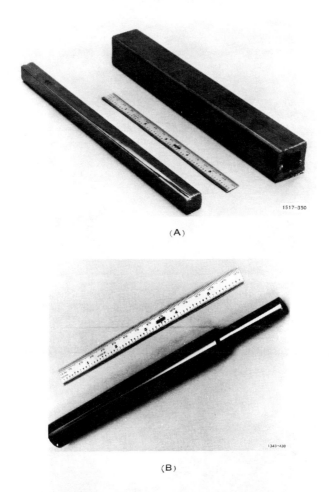

Figure 60. Preforms: (A) Square Extruded, (B) Cast Cylindrical (Reference 15)

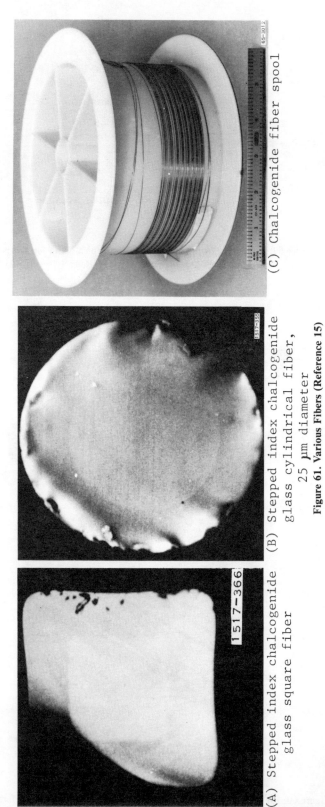

(A) Stepped index chalcogenide glass square fiber

(B) Stepped index chalcogenide glass cylindrical fiber, 25 μm diameter

(C) Chalcogenide fiber spool

Figure 61. Various Fibers (Reference 15)

2.2.4.2 Fiber Drawing

The alternative to the preform technique for fiber fabrication is the drawing of the molten chalcogenide glass directly from a crucible. This has been done with both single and double crucible methods. Figures 62 through 64 are diagrams of these crucible fiber drawing techniques. In either preform or crucible techniques, atmosphere control must be maintained around the hot zone to prevent contamination by ambient oxygen or water. The hot zone is usually kept short to prevent devitrification.

The core and clad glass compositions are selected so the fiber will have some specified NA. Since the NA is determined from the different indices of refraction of the two glasses, the other physical properties, mechanical and thermal, of the core and clad glasses are also different. This can complicate fiber drawing but can also result in mechanically improved fibers. If the elastic moduli, T_g, thermal expansion, and derivative of viscosity with respect to temperature of the two glasses differ greatly, fiber drawing will be uncontrolled or even impossible. If the T_g of the core is lower than the clad and the thermal expansion greater, then the resulting fiber clad will be under a compressive stress. This enhances the strength of the fiber since this stress will be added to the strength of the unstressed fiber. Prestressing the clad can only be done if the two glasses are sufficiently compatible to be drawn into a fiber as discussed previously.

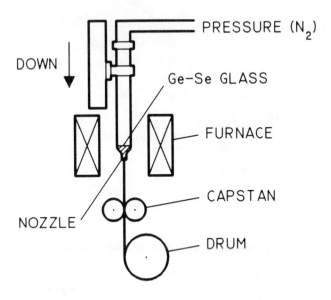

Figure 62. Drawing Apparatus of Chalcogenide Glass Fibers (Reference 7)

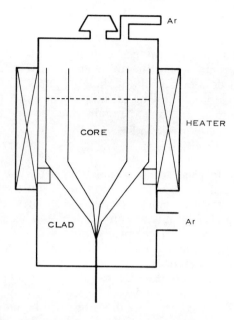

Figure 63. Pyrex Glass Double Crucible Assembly for Preparing Arsenic-Sulfur Glass Fibers (Reference 19)

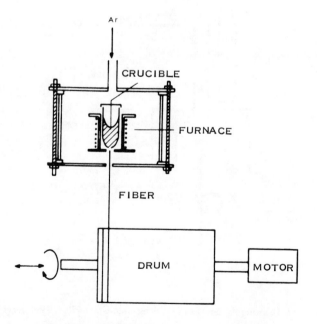

Figure 64. Fiber Pulling Setup (Reference 20)

2.2.5 Waveguide Properties

Figures 65 through 68 show the theoretical attenuation "V" curves and material dispersion curves for some chalcogenide glasses. The extrapolation of the multiphonon edge in these figures is only an estimate using conventional knowledge, whereby one extrapolates from the tail of the reststrahlen band through the multiphonon bands. Table 18 compares the material dispersion zeros with the attenuation minima in the glasses. Figures 69 through 75 are attenuation curves for various chalcogenide glass fibers, obtained from measurements taken with either blackbody or laser sources[29] versus wavelength by means of a monochromator. The chalcogenide glass fibers are orders of magnitude from their theoretical attenuation minima, primarily because of vibrational absorption bands from anion impurities and electronic absorption from metal impurities and structural defects.[29] The difference between the attenuation curves for the fibers and the bulk glasses shown in Figures 51 through 58 is usually attributed to the longer optical path lengths of the fibers. Generally, little contamination is associated with the fiber drawing process other than some surface or interface contamination, as discussed in the preform subsection. Table 19 is a list of the laser power inputs and outputs that have been achieved in some of the chalcogenide fibers. Limited amounts of data have been reported on other physical properties of the fibers such as their thermal, chemical, or mechanical properties. In general, these properties of the fibers are expected to remain the same as the bulk glass. The most important exception to this is the strength of the fibers. Since these fibers are brittle, they fracture because of crack propagation initiated at a flaw in the material. The strength of the fiber will be proportional to the fracture toughness of the glass and inversely proportional to the square root of the flaw size. Current strength values of chalcogenide glass fibers range from a few thousand psi to tens of thousands of psi. Strength values should improve with better fiber fabrication techniques. Table 20 summarizes some strength values for chalcogenide glass optical fibers.

Fabrication techniques for chalcogenide glass fiber optic bundles[36] include redrawing of fibers that have been fused together and epoxying fibers together. Figure 76 is a diagram of this process. Figures 77, 78, and 79 show some of the IR fiber optic bundles that have been fabricated. Improvement in the packing and geometry of the fibers is required to obtain a high quality image bundle.

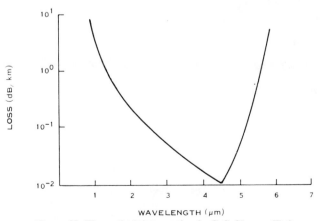

Figure 65. Theoretical Attenuation in GeS Glasses (Reference 21)

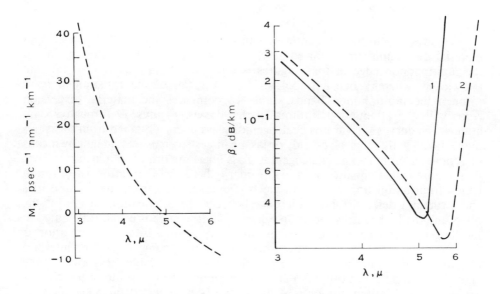

Figure 66. Spectral Dependence of Material Dispersion $M(\lambda)$ of Glassy As_2S_3 and Region of Minimum Optical Losses in Glassy As_2S_3 (1) and As_2Se_3 (2) (Reference 22)

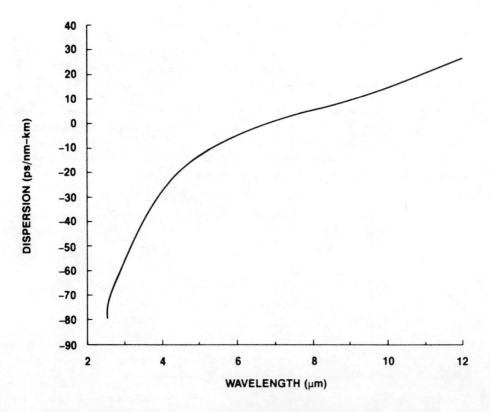

Figure 67. Material Dispersion Versus Wavelength for $Ge_{28}Sb_{12}Se_{60}$ (Reference 10)

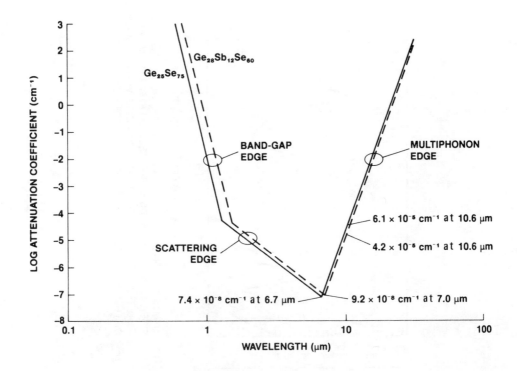

Figure 68. Intrinsic Attenuation Coefficient Versus Wavelength for
$Ge_{25}Se_{75}$ and $Ge_{28}Sb_{12}Se_{60}$ (Reference 10)

TABLE 18. COMPARISON OF ATTENUATION MINIMA AND
MATERIAL DISPERSION ZEROS IN CHALCOGENIDE GLASSES

Composition (Atomic percent)	Attenuation Minima (dB/km)	Material Dispersion Zero Wavelength (μm)	References
$Ge_{25}Se_{75}$	0.03 at 6.7 μm	5.9	10
$Ge_{28}Sb_{12}Se_{60}$	0.04 at 7.0 μm	6.7	10
$As_{40}S_{60}$	0.04 at 5.2 μm	4.9	22
$Ge_{34}S_{66}$	0.04 at 4.7 μm	4.1	23
$As_{40}Se_{60}$	0.08 at 5.0 μm	7.6	23

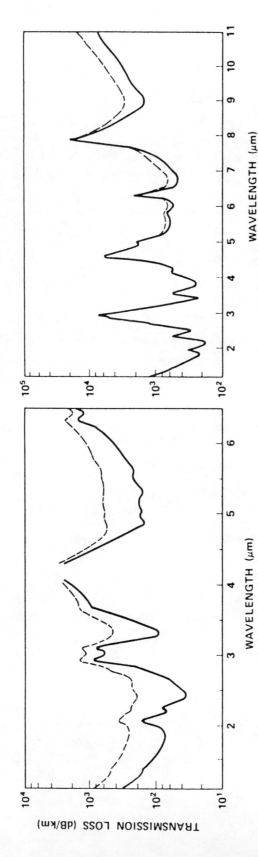

(A) TRANSMISSION LOSS SPECTRA FOR AN $As_{40}S_{60}$ GLASS UNCLAD FIBER 200-μm IN DIAMETER (SOLID LINE) AND As-S GLASS CORE-CLAD FIBER WITH A 120-μm CORE DIAMETER, 200-μm CLADDING DIAMETER, AND A $\Delta\eta$ OF 2.3 PERCENT (BROKEN LINE)

(B) TRANSMISSION LOSS SPECTRA FOR AN $As_{38}Ge_5Se_{57}$ GLASS UNCLAD FIBER WITH A 200-μm DIAMETER (SOLID LINE) AND AN $As_{38}Ge_5Se_{57}$ GLASS CORE-TEFLON FEP CLADDING FIBER WITH A 200-μm CORE DIAMETER AND 220-μm CLADDING DIAMETER (BROKEN LINE)

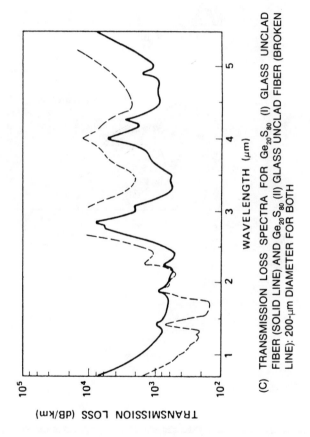

(C) TRANSMISSION LOSS SPECTRA FOR $Ge_{20}S_{80}$ (I) GLASS UNCLAD FIBER (SOLID LINE) AND $Ge_{20}S_{80}$ (II) GLASS UNCLAD FIBER (BROKEN LINE); 200-μm DIAMETER FOR BOTH

Figure 69. Chalcogenide Glass Fiber Attenuation (Reference 24)

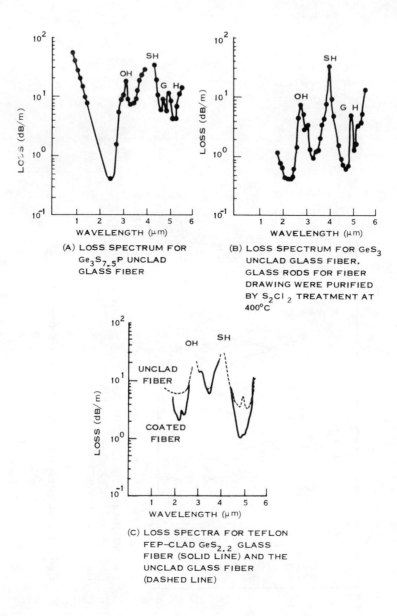

Figure 70. Chalcogenide Glass Fiber Attenuation (Reference 12)

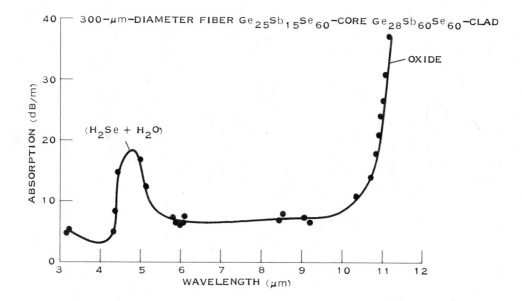

Figure 71. IR Fiber Absorption Versus Wavelength (Reference 15)

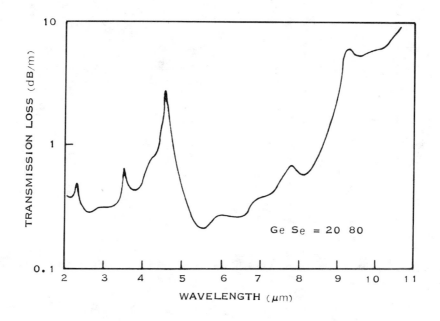

Figure 72. Transmission Loss Spectrum for $Ge_{20}Se_{80}$ Chalcogenide Glass Fiber (Reference 7)

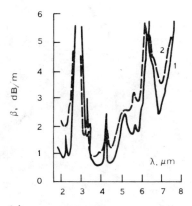

(A) OPTICAL LOSS SPECTRA
1-As_2Se_3 FIBER ($\phi = 530$ μm)
2-WAVEGUIDE ($\phi = 720$ μm)
As_2Se_3 CORE ($\phi = 530$ μm), As_2S_3 CLADDING

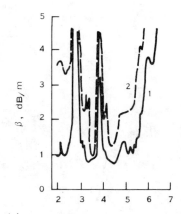

(B) OPTICAL LOSS SPECTRA
1-As_2S_3 FIBER ($\phi = 530$ μm) WITH POLYMER ϕ 42 CLADDING
2-WAVEGUIDE ($\phi = 830$ μm)
As_2S_3 CORE ($\phi = 520$ μm), $As_{33}S_{67}$ CLADDING

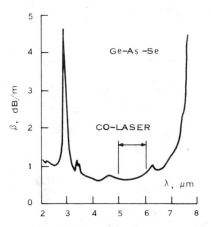

(C) OPTICAL LOSS SPECTRUM OF Ge-As-Se GLASS FIBER WITH TEFLON ϕ42 CLADDING

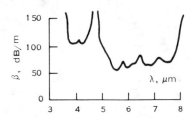

(D) OPTICAL LOSS SPECTRUM OF As-Se FIBER FROM THE GLASS PREFORM PREPARED BY PLASMA CHEMICAL DEPOSITION

Figure 73. Chalcogenide Glass Fiber Attenuation, ϕ = Diameter (Reference 25)

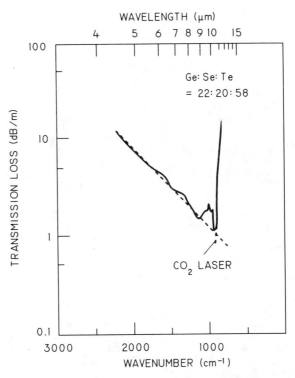

Figure 74. Transmission Loss Spectrum Around 10.6 μm for a $Ge_{22}Se_{20}Te_{58}$ Glass Optical Fiber (Reference 26)

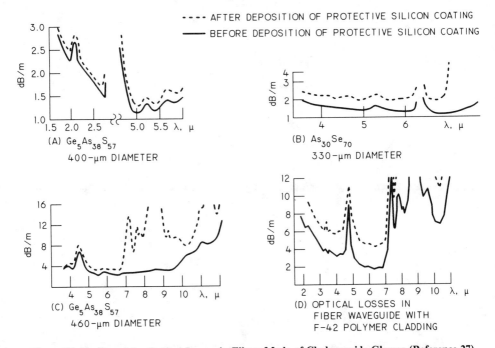

Figure 75. Spectra of the Optical Losses in Fibers Made of Chalcogenide Glasses (Reference 27)

TABLE 19. LASER POWER TRANSFER IN CHALCOGENIDE GLASS OPTICAL FIBERS

Fiber Type	Laser	Power In (W)	Power Out (W)
As_2Se_3 (unclad) (Reference 30) (0.2-meter length, 200-μm diameter)	CO_2	3	0.25
$Ge_5As_{38}Se_{57}$ (Teflon clad) (Reference 25) (1-meter length, 500- to 700-μm diameter)	CO	10 to 12	6 to 7
As_2S_3 (unclad) (Reference 31) (4-meter length, 1,000-μm diameter)	CO	Unknown	40
As_2S_3 (FEP clad) (Reference 32) (0.55-meter length, 700-μm diameter)	CO	100	62

TABLE 20. STRENGTH OF CHALCOGENIDE GLASS OPTICAL FIBERS

Fiber Type	Test Type	Strength (MPa)
150-μm AMTIR glass, unclad (Reference 33)	Tensile Bend	81 to 114 208 to 430
300-μm AMTIR glass, unclad (Reference 33)	Bend	146
150-μm AMTIR core/clad glass (Reference 33)	Bend	310 to 489
94-μm $Ge_5As_{38}Se_{57}$ (Reference 34)	Tensile	177 ±40
82-μm $Ge_5As_{38}Te_{57}$ plastic clad (Reference 34)	Tensile	285 ±25
700-μm As_2S_3 Teflon clad (Reference 35)	Tensile	50 to 100

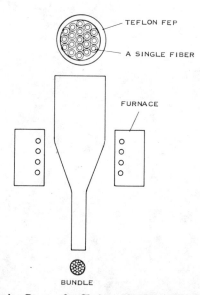

Figure 76. Drawing Process for Chalcogenide Fiber Bundles (Reference 37)

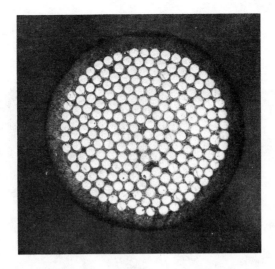

Figure 77. Cross-Sectional Picture of IR Fiber Bundle, 2.0 mm in Diameter Including 200 AsS Glass Fiber Cores, Each 90 μm in Diameter (Reference 37)

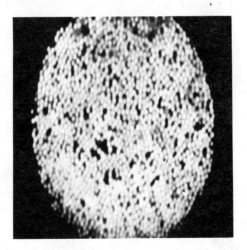

Figure 78. Coherent Image Bundle, 25-μm-Diameter Chalcogenide Glass Fibers, Shown in a Transmission IR Microscope (Reference 15)

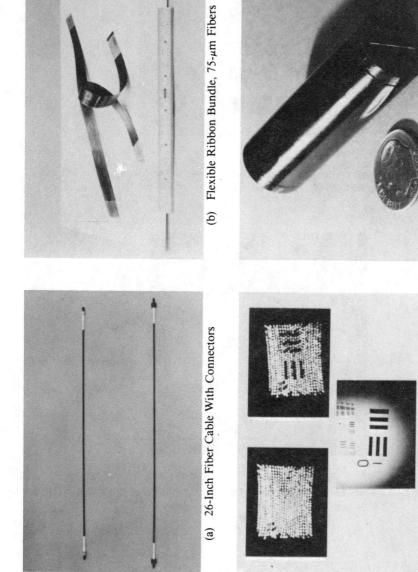

(a) 26-Inch Fiber Cable With Connectors
(b) Flexible Ribbon Bundle, 75-μm Fibers
(c) Coherent Image Bundle, 300-μm Fiber
(d) Line-to-Spot Reformatter, 75-μm Fiber

Figure 79. IR Fiber Optic Prototypes (Reference 15)

2.3 CRYSTALLINE MATERIALS

2.3.1 Composition

Crystalline materials have been used for many years as IR optical components. Generally, the metal halides, Group VII elements, which have very wide transmission from the UV well into the IR, have been fabricated into optical fibers. Table 21 lists both the single and polycrystalline compositions of those materials used to fabricate IR fibers. Table 21 includes single crystal oxide compositions that have been formed into optical fibers. The oxides, however, have a transparency limited to 4 or 5 μm.

TABLE 21. MATERIALS FOR INFRARED FIBERS

KRS–5 (TlBrI)	Polycrystalline and single crystal
TlBr	Polycrystalline
AgCl	Polycrystalline
AgBr	Polycrystalline and single crystal
KCl	Polycrystalline
CsBr	Single crystal
CsI_2	Single crystal
BaF_2	Single crystal
CaF_2	Single crystal
ZnSe	Polycrystalline
Al_2O_3	Single crystal
ADP	Single crystal
KDP	Single crystal
$NaOH_2$	Single crystal
CuCl	Single crystal

2.3.2 Material Preparation

Metal halide and oxide crystals are usually prepared by crystal growth from a molten solution of the elements. High purity and proper temperature control are essential. A solid crystal can be obtained by cooling the entire melt or by placing a seed in the melt and pulling it as the crystal grows out of the melt. This latter technique is called the Czochralski method. A Bridgman technique can also be used whereby a molten solution of the elements is quenched from one end to the other by translating the melt through a thermal gradient where one end of the gradient is above the crystal melting point and the other end is below. Traveling heater method (THM) has also been used where a narrow heat zone above the melting temperature of the crystal is translated from one end of a solid crystal to the other end. A solvent may be introduced that lowers the melting point of the crystal. As the hot zone moves, it regrows the crystal usually as a single crystal with low dislocations and defects and higher purity owing to the segregation of impurities. Similar considerations about impurities causing both electronic and vibrational extrinsic absorption that were discussed in the subsections on glasses also apply to crystalline materials. An additional problem with crystalline materials, however, is the presence of crystal defects and grain boundaries that give rise to extrinsic absorption and scattering. Further details on crystal growth[1] can be found in many references.

2.3.3 Physical Properties

Halide crystals, unlike the oxide crystals, offer the lowest theoretical attenuation of any material, primarily because of the steep exponential behavior of their multiphonon edge and low theoretical light scattering. Achievement of these theoretical properties, however, is far from being realized. Table 22 shows the IR transmission range of the halide crystals as well as the oxides. Table 23 lists the absorption coefficients for many crystals. Table 24 lists the physical properties of the crystals. Oxide crystals are of interest as IR optical fibers, primarily for nonlinear optical applications.

TABLE 22. IR TRANSMISSION IN CRYSTALLINE MATERIALS FOR FIBERS

Crystal	Transmission Range (μm) 10-Percent Cutoffs, 2 mm Thick
KRS–5	0.6 to 40.0
TlBr	0.42 to 48.0
AgCl	0.4 to 28.0
AgBr	0.45 to 40.0
KCl	0.21 to 30.0
CsBr	0.3 to 55.0
CsI	0.25 to 80.0
BaF_2	0.25 to 15.0
CaF_2	0.13 to 12.0
ZnSe	0.5 to 20.0
Al_2O_3	0.2 to 6.0
ADP	0.1 to 2.0
KDP	0.2 to 2.0

TABLE 23. ABSORPTION COEFFICIENTS OF IR CRYSTALLINE MATERIALS FOR FIBERS

Crystal	Present Bulk Absorption (cm^{-1})
KRS–5	7×10^{-4} at 10.6 μm
TlBr	1×10^{-3} at 10.6 μm
AgCl	5×10^{-3} at 10.6 μm
AgBr	5×10^{-3} at 10.6 μm
KCl	8×10^{-5} at 10.6 μm
CsBr	$<10^{-3}$ at 10.6 μm
CsI	$<10^{-3}$ at 10.6 μm
BaF_2	$<10^{-3}$ at 4 μm
CaF_2	1.7×10^{-4} at 3.8 μm
ZnSe	5×10^{-4} at 10.6 μm
Al_2O_3	1×10^{-4} at 4 μm

TABLE 24. PHYSICAL PROPERTIES OF IR CRYSTALLINE MATERIALS FOR FIBERS (REFERENCE 2)

Cyrstal	Melting Point (°C)	Thermal Expansion ($\times 10^{-6}$/°C)	Knoop Hardness (kg/mm²)	Young's Modulus ($\times 10^6$ psi)	Index of Refraction at 5 μm	Solubility (g/100 g H_2O)
KRS–5	414	61.0	38.0	2.3	2.38	5.0×10^{-2}
TlBr	460	51.0	11.9	4.3	2.35	4.8×10^{-2}
AgCl	455	30.0	9.5	0.02	1.96	1.5×10^{-4}
AgBr	432	35.0	9.5	0.02	2.00	8.4×10^{-6}
KCl	790	36.0	8.0	4.3	1.46	34.3
CsBr	636	46.0	19.5	2.3	1.66	124.3
CsI	621	50.0	—	0.77	1.72	—
BaF_2	1,355	18.4	82.0	7.7	1.46	0.17
CaF_2	1,360	18.4	163.0	11.0	1.40	1.7×10^{-3}
ZnSe*	1,520	8.0	150.0	6.0	2.43	Insoluble
Al_2O_3**	2,040	5.3	2,000.0	50.0	1.61	Insoluble
ADP (Reference 3)	190	39.3	—	—	1.50 at 1.1 μm	22.7
KDP (Reference 3)	253	42.0	—	—	1.49 at 1.1 μm	33.0
NaOH	307	12.0	20.0	—	1.57 at 0.7 μm	100.0

*Data from Raytheon Inc.
**Data from Crystal Systems.

2.3.4 Waveguide Fabrication

Fabrication of waveguides from crystalline materials is significantly more difficult than with glass materials. This is caused by the lack of any viscoelastic properties in crystalline materials where there is a single temperature at which they transition from liquid to solid. While high-quality IR crystals have been fabricated, it is more difficult to maintain that quality while transforming that crystal into an optical fiber. Although it is a reasonably simple task to clad a glass optical fiber as described in previous subsections, it is very difficult to clad a crystalline fiber. Cladding is necessary to have a controlled NA and to minimize attenuation by protecting the outer surface of the fiber core and providing a low-loss medium for the evanescent field of the energy in the fiber.

Several methods have been used to fabricate IR optical fibers from crystalline materials. One is an extrusion method of a crystal rod where solid-state recrystallization and plastic deformation occurs through a die and forms the fiber. Fibers formed by this technique are polycrystalline. Figure 80 shows this technique. Another method, shown in Figure 81, involves fiber growth from the liquid phase as it is forced through a nozzle. A third technique is called the laser pedestal growth method, in which a crystal rod is heated with power from a laser and, with exact thermal control, recrystallization of the melt zone results in a single crystal fiber. This technique allows a great variety of crystalline materials to be fabricated into fibers that would be difficult or impossible with the other techniques. For instance, incongruently melting crystals, doped crystals, and solid solution crystals could be made into optical fibers using the laser pedestal method. Figure 82 is a diagram of the laser pedestal growth method. Figure 83 is a diagram of modified Bridgman, THM, and Czochralski methods used to provide crystalline fibers. Cladded fibers have been attempted by coextrusion processes where a crystal rod is placed inside a bored-out crystal.[8] Table 25 lists

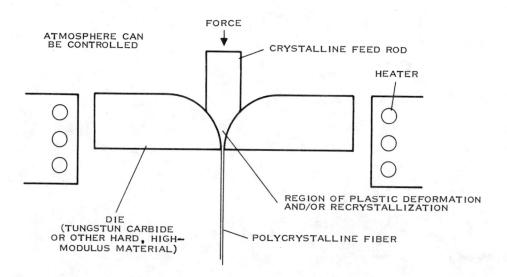

Figure 80. Crystalline Fiber Extrusion

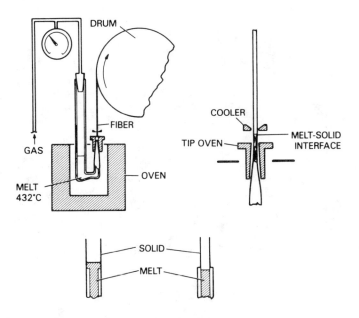

Figure 81. Schematic of Fiber Crystal-Growing Apparatus (Reference 5)

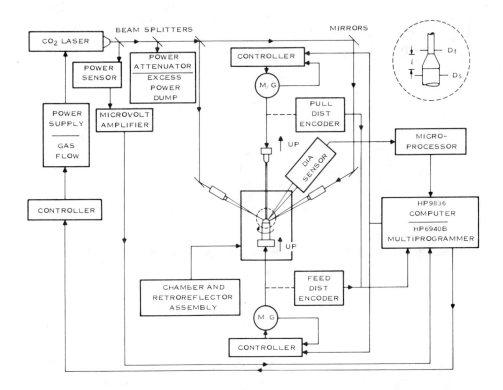

Figure 82. Schematic Diagram of Laser-Heated Pedestal Growth System (Reference 6)

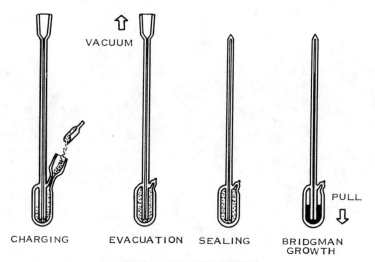

Figure 83. Continued

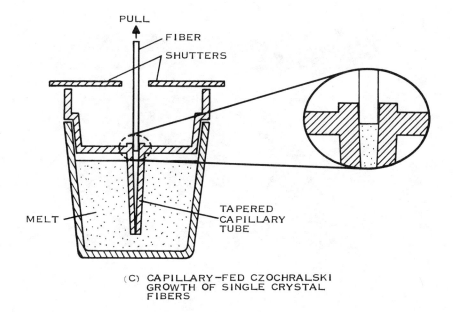

(C) CAPILLARY-FED CZOCHRALSKI GROWTH OF SINGLE CRYSTAL FIBERS

Figure 83. Crystal Fiber Growth (Reference 7)

TABLE 25. FABRICATION PARAMETERS OF IR CRYSTALLINE FIBERS

Reference	Crystal	Fabrication Technique	Fiber Fabrication Rate (in./min)	Fiber Diameter (μm)	Temperature (°C)	Grain Size (μm)	Extrusion Pressure (psi)
3	AgBr	Extrusion	0.200–25	75–450	20–310	1–50	$\sim 2.5 \times 10^5$
4	KRS-5	Extrusion	0.200–25	75–10,000	200–350	3–50	$\sim 2.5 \times 10^5$
4	TlBr	Extrusion	0.200–25	75–10,000	200–350	3–50	$\sim 2.5 \times 10^5$
4	KCl	Extrusion	0.200–5	500–1,000	200–350	3–50	$\sim 2.5 \times 10^5$
9	CsBr	Melt	0.400	800	~ 650	Single	—
5	BaF$_2$	Pedestal	0.001–4	200–600	~ 1300	Single	—
5	CaF$_2$	Pedestal	0.001–4	600	~ 1400	Single	—
10	Al$_2$O$_3$	Pedestal	0.04–0.4	150	$\sim 2,000$	Single	—
6	ADP	Bridgman	0.001	500	~ 200	Single	—

some of the parameters involved in the different crystalline fiber fabrication methods. Figures 84, 85, and 86 show crystalline fibers produced by some of these methods. Crystalline fibers usually suffer from rough surfaces that increase the light scattering in the fiber and reduce its strength. Traveling heater[12] recrystallization into single crystal fibers and annealing[13] techniques have been used to improve the surface quality of the fibers.

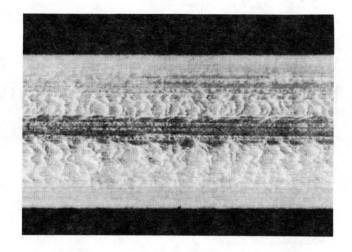

(A) 8X, EXTRUDED KCl, MEDIUM RATE, 0.25 IN DIAMETER

(B) 8X, EXTRUDED KCl, HIGH RATE, 0.25 IN DIAMETER

Figure 84. Extruded KCl, 0.25 Inch in Diameter, 8× (Reference 4)

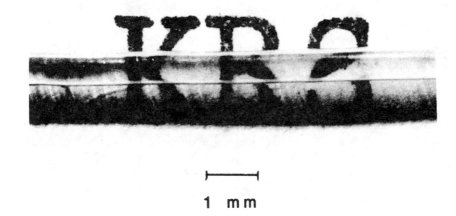

Figure 85. KRS-5 Fiber From Laser Pedestal Growth Method (Reference 11)

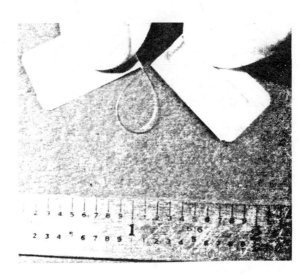

Figure 86. Demonstration of Flexibility of a 150-μm C-Axis Sapphire Fiber
(Ruler is Marked in Inches) (Reference 10)

2.3.5 Waveguide Properties

Figures 87 through 89 show the theoretical attenuation "V" curves for some crystalline IR fiber materials. Table 26 lists the material dispersion zeros for the crystalline materials and compares them with the wavelength at which the attenuation minimum occurs. Attenuation versus wavelength for crystalline fibers fabricated by various methods appears in Figures 90, 91, and 92. Light scattering is usually responsible for much of the attenuation in these fibers. Physical property data for the various fibers is compiled in Table 27, as is the laser power handling capacity of some of the crystalline fibers. The polycrystalline fibers suffer from extrinsic scattering from residual strain in the crystals and the surface irregularities at the fiber surface.[23] The polycrystalline fibers also tend to "age"; that is, there is a degradation of their optical and mechanical properties over time. This appears to be the result of relaxation of residual stress in the fibers. Many crystal fibers also degrade optically and remain deformed after bending owing to plastic deformation of the crystals.

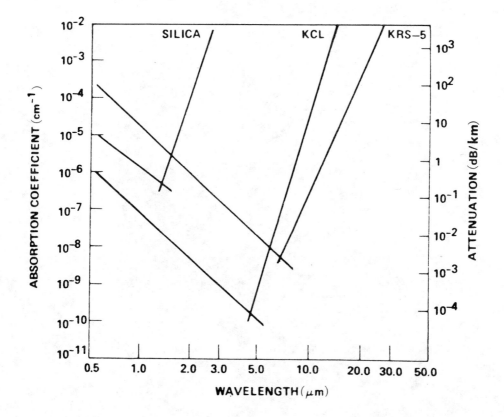

Figure 87. Projected Transmission in IR Fibers (Reference 14)

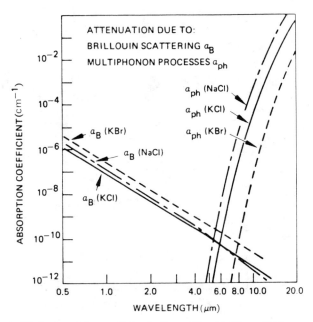

Figure 88. Projected Transmission in Crystalline IR Fibers (Reference 15)

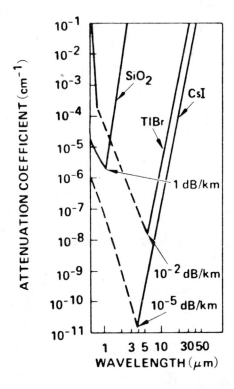

Figure 89. Projected Transmission Loss in IR Fibers (Reference 15)

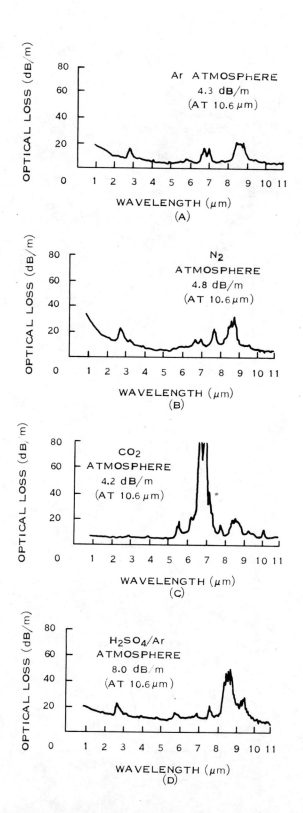

Figure 90. IR Loss Spectra for CaBr Fibers Grown in Different Atmospheres (Reference 9)

TABLE 26. MATERIAL DISPERSION ZEROS AND THEORETICAL ATTENUATION MINIMUM IN CRYSTALLINE IR FIBERS

Crystal	Material Dispersion Zero (μm) (Reference 15)	Theoretical Attenuation Minimum (dB/km)
AgCl	5.07	—
KCl	3.10	7×10^{-5} at 5.9 μm (Reference 16)
AgBr	6.80	—
TlBr	8.50	1×10^{-2} at ~5.0 μm (Reference 15)
KRS-5	~7.50	4×10^{-3} at ~7.0 μm (Reference 14)
CsBr	5.40	—
CsI	6.80	1×10^{-5} at ~4.0 μm (Reference 15)
BaF$_2$	2.40	—
CaF$_2$	1.60	—
ZnSe	5.50	—

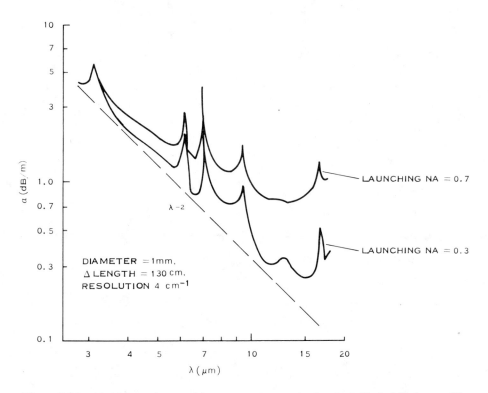

Figure 91. Total Loss Spectrum of KRS-5 Fiber, Measured by Cut-Back Method (Reference 17)

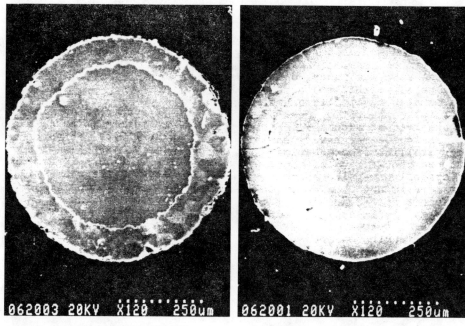

(A) EXTRUDED STEP INDEX TYPE FIBER

(B) GRADED INDEX TYPE FIBER AFTER HEAT TREATMENT

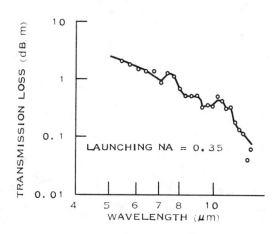

Figure 92. Extruded Crystalline Fiber of KRS-5/KRS-6 Cone/Clad (Reference 8)

TABLE 27. ATTENUATION IN CRYSTALLINE IR FIBERS

Crystal	Fiber Type	Fabrication Method	Attenuation at 10.6 μm (dB/m)	CO_2 Laser Output Power (W)
Poly KRS–5 (Reference 18)	Teflon sheath, 500-μm diameter	Extrusion	0.46	2
Poly KRS–5 (Reference 14)	Unclad, 500-μm diameter, 1 m long	Extrusion	0.30	9
Poly KRS–5 (Reference 19)	Unclad, 1000-μm diameter	Extrusion	0.40	97
Single AgBr (Reference 3)	Unclad, 750-μm diameter	Melt	8.70	4
Poly TlBr (Reference 13)	Unclad, 500-μm diameter, 1 mm long	Extrusion	0.43	—
Poly KCl (Reference 13)	Unclad, 500-μm diameter, 1 m long	Extrusion	4.20	—
Poly AgCl (Reference 14)	Unclad, 500-μm diameter	Extrusion	4.00	—
Single CsI (Reference 20)	Teflon sheath, 700-μm diameter, 1.5 m long	Melt	80.00	—
Single CsBr (Reference 20)	Teflon sheath, 700-μm diameter, 1.5 m long	Melt	5.00	—
Poly AgCl (Reference 21)	Unclad, 450-μm diameter	Extrusion	6.00 at 14 μm	—
KRS–5/KRS–6 (Reference 8)	Clad, 700-μm diameter	Coextrusion	0.2	—
Al_2O_3 (Reference 10)	Unclad, 150-μm diameter	Pedestal	0.070 at 3.39 μm	—
Silver Halide (Reference 22)	Teflon sheath, 500- to 1000-μm diameter, 1 to 2 m long	Extrusion	0.5	20 to 50

2.4 IR OXIDE GLASSES

There are some heavy metal oxide glasses that offer some IR transparency out to 5 μm. Table 28 lists these glass compositions and some of their physical properties. Figure 93 shows the index of refraction for each of the compositions. Figure 94 shows the IR spectra of some GeO_2-Sb_2O_3 glasses. Figure 95 shows the schematic of a CVD process that has been reported for the production of IR fibers from these heavy metal oxide glasses. The theoretical attenuation "V" curve for the GeO_2-Sb_2O_3 glass is shown in Figure 96. Table 29 lists the material dispersion zeros for these glasses. The results for coated and uncoated fiber are shown in Figure 97.

TABLE 28. COMPOSITION AND SOME PROPERTIES OF IR OXIDE GLASSES (REFERENCE 1)

Glass	A	B	C	D	E
Composition					
GeO_2	30	35	41	40	50
$BiO_{1.5}$	30	12	—	—	20
$TlO_{0.5}$	40	41	39	—	—
PbO	—	12	20	20	30
$SbO_{1.5}$	—	—	—	40	—
Annealing temperature (°C)	260	200	200	325	400
Annealing time (h)	3	3	3	1	16
Density (g cm^{-3})	7.66	7.35	6.98	5.40	5.99
Thermal expansion (10^{-6} K^{-1})	15.00	14.20	15.90	13.00	10.60
Glass transition temperature (°C)	270	260	240	350	470
Crystallization temperature (°C)	370, 430	380	400	—	—

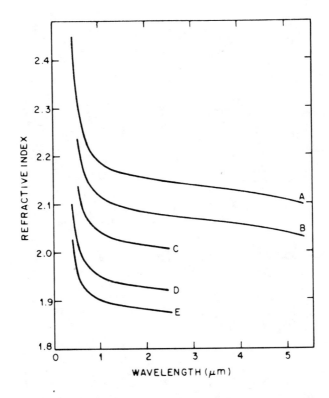

Figure 93. Refractive Index Variation With Wavelength for Five Glasses of Table 29 (Reference 1)

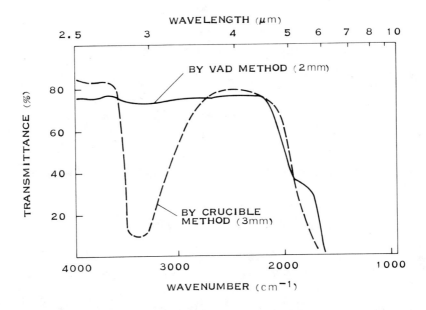

Figure 94. IR Transmission Spectrum of Germanate Glasses Prepared By Crucible Method and by VAD Method (Reference 2)

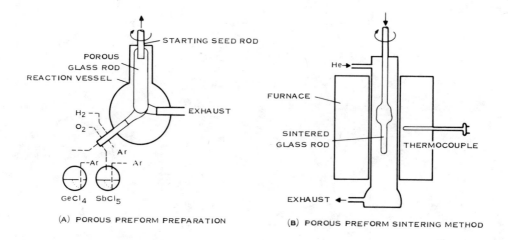

Figure 95. Schematics of Porous Preform Preparation and Sintering Method (Reference 2)

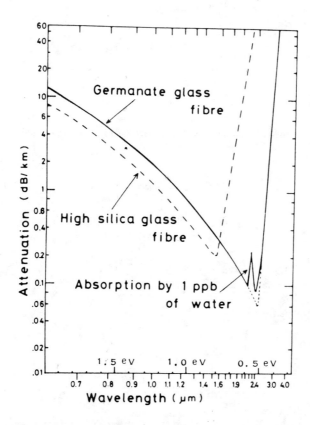

Figure 96. Calculated Theoretical Loss Spectrum of Germanate Glass Fiber (Reference 2)

TABLE 29. OPTICAL PROPERTIES OF GLASSES IN TABLE 28 (REFERENCE 1)

Glass	A	B	C	D	E
n_D	2.28553	2.19901	2.11322	1.99561	1.93948
$n\lambda_o$	2.14175	2.07484	2.00648	1.92566	1.88182
Disp. = $n_F - n_c$	0.12226	0.10379	0.08576	0.05115	0.04137
Abbe number	10.51	11.55	12.98	19.46	22.71
Observed λ_o (μm)	2.81	2.73	2.65	2.22	2.08
Predicted λ_o (μm)	3.3	3.1	3.0	2.4	2.5

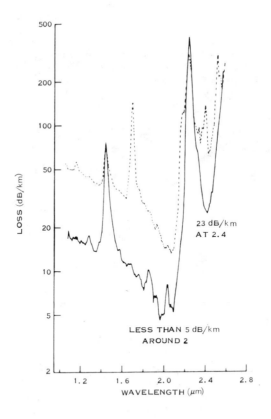

Figure 97. Measured Loss Spectrum of Germanate Glass Fibers With Silicone Resin Coating (Dashed Line) and Without Any Coating (Solid Line) (Reference 2)

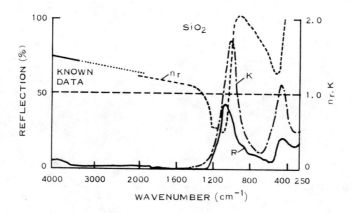

REFLECTION COEFFICIENT R AND COMPLEX REFRACTIVE INDEX $n_1 = n_r - ik$ OF FUSED SiO_2. THE SCALE FOR R (SOLID LINE) IS ON THE LEFT ORDINATE, AND THE SCALE FOR n_r (DASH LINE) AND K (DOT DASH LINE) IS ON THE RIGHT. n_r IS SMALLER THAN UNITY IN THE FREQUENCY BETWEEN 1040 AND 1300 cm^{-1}. THE VALUE OF n_r EXTRAPOLATED FROM THE KNOWN DATA (ABOVE 3000 cm^{-1}) TO THE FREQUENCY LESS THAN 2000 cm^{-1} IS IN GOOD AGREEMENT WITH OUR DATA. THE MINIMUM n_r IS 0.47 AT 1090 cm^{-1}.

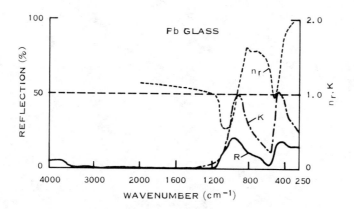

R AND $n_1 = n_r - ik$ FOR Pb SILICATE GLASS. n_r IS SMALLER THAN UNITY FROM 930 TO 1210 cm^{-1} AS WELL AS FROM 500 TO 540 cm^{-1}

Figure 98. Complex Index and Reflection Coefficient for SiO_2 and Pb-Glass (Reference 1)

2.5 HOLLOW WAVEGUIDES

An example of a cylindrical dielectric hollow IR waveguide is shown in Figures 98 and 99. Figure 98 presents the reflection coefficient for silica and a lead silicate glass. Figure 99 shows the transmission of a lead silicate glass hollow waveguide. Figure 100 shows the transmission results for a silver metal cylindrical hollow waveguide. One technique used to reduce the bending losses described in Section 1 is to deposit transparent dielectric layers inside the metal waveguide, which impedance matches the waveguide walls. Figure 101 shows some calculated and measured results for these layers.

Fabrication of hollow metal waveguides of lengths greater than a meter have not been reported. Internal surfaces must be highly polished to minimize scattering. The surfaces can quickly increase attenuation in the waveguide if not maintained in a highly polished form. The difficulty in bending a metal, which is usually stiff or will plastically deform, and the dimensions of these waveguides are serious limitations. They can, however, carry high laser powers without damage if a laser is available with the proper modal output at high powers. Table 30 summarizes some hollow waveguide transmission results.

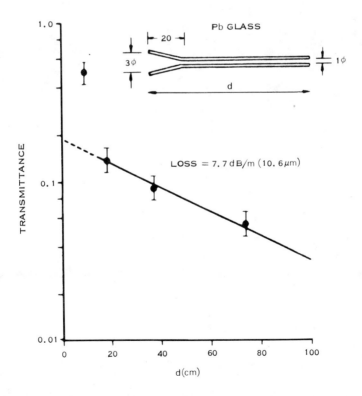

Figure 99. Experimental Result for Transmission of Pb-Glass Hollow-Core Fiber for CO_2 Laser Light (940 cm^{-1}) (Inner diameter is 1.0 mm; measured loss is for multimode transmission) (Reference 1)

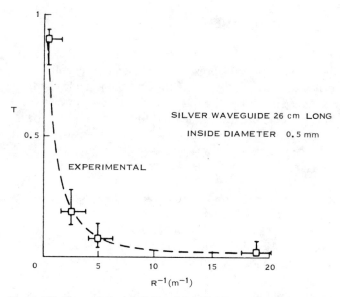

Figure 100. Transmission of Hollow Metallic Waveguide of Cylindrical Cross Section as Function of Inverse Bend Radius (Reference 2)

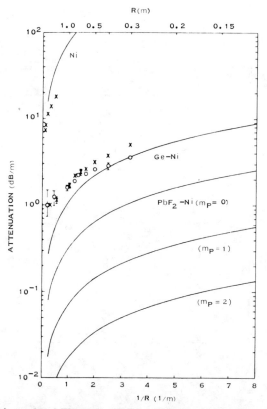

Figure 101. Experimental and Theoretical Bending Losses of Nickel Cylindrical Waveguides Uncoated and Coated by Ge and PbF_2, where X and O Correspond to Measured $E_\parallel$ and $E_\perp$, Respectively (Reference 3)

TABLE 30. HOLLOW WAVEGUIDE TRANSMISSION

Waveguide Type	Reference	Size (mm)	Length (cm)	Attenuation Transmission	CO_2 Laser Power Output (W)
Straight dielectric (Pb silicate glass), cylindrical	1	1 (dia.)	100	7.7 dB/m at 10.6 μm	—
Straight metal (smooth Ag), cylindrical	2	0.5 (dia.)	31	90% at 10.6 μm	—
Straight metal (commercial copper and stainless steel), cylindrical	2	0.5 (dia.)	17	78 and 55% at 10.6 μm	—
Bent (90°, 10 cm radius) metal, rectangular	2	1 × 30	1,400	70% at 10.6 μm	200
Straight metal (Au coated stainless steel), rectangular	4	0.5 × 10	30	96% per meter at 10.6 μm	840
Bent (90°, 20 cm radius) metal (Au coated stainless steel), rectangular	4	0.5 × 10	100	92% per meter at 10.6 μm	770
Straight dielectric (GeO_2 glass), cylindrical	5	1 (dia.)	1	3.5 dB/m at 10.6 μm	—
Bent dielectric (GeO_2 glass) (30 cm radius), cylindrical	5	1 (dia.)	1	11.5 dB/m at 10.6 μm	—

2.6 SUMMARY

Table 31 presents a brief summary of the various advantages and disadvantages of IR optical fibers. Discussions about the reasons for the classifications in the table are found in the appropriate sections of this book. Selection of one or more of these fiber types will depend on the specifications required. Table 31 is a general guide to the current state of the technology. Note that this table indicates IR fiber technology has progressed to the point of practical use. However, only a thorough knowledge of each IR fiber type and its current level of development will lead to the proper choice for a specific application.

Finally, an interesting IR fiber development that has not been previously mentioned is a liquid-filled flexible hollow tube.[1,2] Liquid bromine has been inserted into a Teflon tube (1.6 mm in diameter and 1 m long) with attenuation of 0.5 dB/m at 9.6 μm. Bromobenzene has been used inside silicate glass tubes 50 meters long and 125 μm in diameter, achieving 0.14 dB/m at 0.63 μm.[2] Little more has been reported on this type of fiber. Although possibly useful in the IR region, fabrication and toxicity may cause concern.

TABLE 31. SUMMARY OF IR OPTICAL FIBERS

Fiber Type	Material	Transmission Range (μm) (Note 1)	Current Attenuation (dB/m) (Note 2)	Mechanical/ Chemical Durability	Fabrication	Flexibility	Laser Power Handling
Unclad dielectric	Fluoride glass	0.3 to 6.0	0.01 to 50.0	Moderate	Good	Moderate	Good
	Chalcogenide glass	0.8 to 16.0	0.04 to 50.0	Moderate	Good	Moderate	Moderate
	Crystalline	0.3 to 20	0.2 to 50.0	Poor	Moderate	Poor	Good
Clad dielectric	Fluoride glass	0.3 to 6.0	0.001 to 50.0	Moderate	Good	Moderate	Good
	Chalcogenide glass	0.8 to 16.0	0.04 to 50.0	Moderate	Good	Moderate	Moderate
	Crystalline	0.3 to 20	0.2 to 50.0	Poor	Poor	Poor	Good
	Heavy metal oxide	0.4 to 5.0	0.005 to 50.0	Good	Good	Good	Good
Hollow cylindrical	Dielectric	Narrow but selectable	7 to 50	Good	Good	Poor	Moderate
	Metal	Broad	5 to 50	Good	Moderate	Poor	Good
	Metal/dielectric	Broad but limited by dielectric	1 to 50	Good	Moderate	Poor	Good
Hollow rectangular	Metal	Broad	0.2 to 50.0	Good	Moderate	Poor	Excellent
	Metal/dielectric	Broad but limited by dielectric	0.2 to 50.0	Good	Moderate	Poor	Excellent

Notes:

1. Assumes transmission thickness of 1 cm of bulk material, except for hollow waveguides.
2. Lowest value is minimum reported; upper value has been limited to 50 and is highly dependent on wavelength and other parameters.

2.7 REFERENCES

2.7.1 Subsection 2.1

1. T. Miyashita and T. Manabe, "Infrared optical fibers," *IEEE J. Quant. Elec.* **QE-18** (1982), pp. 1432–1450.

2. D.C. Tran, G.H. Sigel, Jr., and B. Bendow, "Heavy metal fluoride glasses and fibers: A Review," *J. Lightwave Technology* **LT-2** (1984), pp. 566–586.

3. M.G. Drexhage, "Heavy Metal Fluoride Glasses," *Treatise on Materials Science*, Vol. 26, ed. by M. Tomozawa and R. Doremus, Academic Press, New York (1985), pp. 151–243.

4. K.H. Sun, "Fluoride glass," U.S. Patent 2,466,509 (1949).

5. S. Takahashi et al., "New fluoride glasses for IR transmission," in *Adv. in Ceramics*, Vol. II of *Physics of Fiber Optics*, ed. by B. Bendow and S.S. Mitra, ACS (1981), p. 74.

6. C. Jacoboni, A. le Bail, and R. de Pape, "Fluoride glasses of 3D transition metals," *Glass Tech.* **24**:3, (1983), pp. 164–167.

7. G. Fonteneau, H. Slim, F. Lahaie, and J. Lucas, "Nouveaux verres fluores transmetteurs dans l'infrarouge dans les systemes LnF_3-BaF_2-ZnF_2," *Mat. Res. Bull.* **15** (1980), pp. 1425–1432.

8. G. Fonteneau, F. Lahaie, and J. Lucas, "Une nouvelle famille de verres fluores transmetteurs dans l'infrarouge: fluorures vitreux dans les systemes ThF_4-BaF_2-MF_2 (M = Mn, Zn)," *Mat. Res. Bull.* **15** (1980), pp. 1143–1147.

9. S. Mitachi, Y. Ohishi, and S. Takahashi, "Preparation of fluoride optical fiber," *Rev. Elec. Commun. Labs* **32**:3 (1984), p. 118.

10. D.C. Tran et al., "Fluoride glass optical fibers," *Mat. Sci. Forum 5* (1985), pp. 339–352.

11. P.W. France, S.F. Carter, S.F. Williams, and K.J. Beales, "OH Absorption in Fluoride Glass Infrared Fibers," *Electronics Lett.* **20** (1984), pp. 607–608.

12. M.G. Drexhage et al., "Progress in the development of multispectral glasses based on the fluorides of heavy metals," *Proceedings of the 13th Boulder Laser Damage Symposium* (1982).

13. D.C. Tran, G.H. Sigel, Jr., K.H. Levin, and R.J. Ginther, "Rayleigh scattering in ZrF_4-based glass," *Electronics Lett.* **18** (1982), pp. 1046–1048.

14. D.C. Tran, K.H. Levin, R.J. Ginther, G.H. Sigel, Jr., and A.J. Bruce, "Light Scattering in heavy-metal fluoride glasses in infrared spectral region," *Electronics Lett.* **22** (1986), pp. 117–118.

15. K.H. Levin, D.C. Tran, R.J. Ginther, and G.H. Sigel, Jr., "Optical properties of fibre and bulk zirconium fluoride glass," *Glass Technol.* **24** (1983), p. 143.

16. S. Shibata, M. Horiguchi, K. Jinguji, S. Mitachi, T. Kanamori, and T. Manabe, "Prediction of loss minima in infrared optical fibres," *Electronics Lett.* **17** (1981), pp. 775–777.

17. B. Bendow, "Fundamental optical properties of heavy metal fluoride glasses," paper 5, Second Int. Symposium on Halide Glasses, New York (1983).

18. K.H. Levin, D.C. Tran, R.J. Ginther, and G.H. Sigel, Jr., "Optical properties of fibre and bulk zirconium fluoride glass," *Glass Technol.* **23** (1983), p. 143.

19. B. Bendow, "Mid-IR fiber optics technology study and assessment," NOSC Final Report (February 28, 1984).

20. D.C. Tran, R.J. Ginther, and G.H. Sigel, Jr., "Fluorozirconate glasses with improved viscosity for fiber drawing," *Mat. Res. Bull.* **17** (1982), pp. 1177–1184.

21. Y. Mimura, H. Tokiwa, and O. Shinbori, "Fabrication of fluoride glass fibres by the improved crucible technique," *Electronics Lett.* **20** (1984), pp. 100–101.

22. S. Mitachi, S. Shibata, and T. Manabe, "Teflon FEP-clad fluoride glass fibres," *Electronics Lett.* **17** (1981), pp. 128–129.

23. M.G. Drexhage, B. Bendow, T.J. Loretz, J. Mansfield, and C.T. Moynihan, "Preparation of multicomponent fluoride glass fibers by the single crucible technique," *Tech. Digest IOOC '81*, Paper M12 (1981).

24. S. Mitachi, T. Miyashita, and T. Kanamori, "Fluoride glass cladded optical fibers for mid-infrared ray transmission" *Electronics Lett.* **17** (1981), pp. 591–592.

25. S. Mitachi, T. Miyashita, and T. Manabe, "Preparation of fluoride optical fibres for transmission in the mid-infrared," *Phys. & Chem. of Glasses* **23** (1982), p. 24.

26. D.C. Tran, C.F. Fisher, and G.H. Sigel, Jr., "Fluoride glass preforms prepared by a rotational casting process," *Electronics Lett.* **18** (1982), pp. 657–658.

27. D.C. Tran, M.J. Burke, G.H. Sigel, Jr., and K.H. Levin, "Preparation of single-mode and multimode graded-index fluoride-glass optical fibers using a reactive vapor transport process," paper TUG2, *Tech. Digest, Conference on Optical Fiber Communication*, New Orleans (1984).

28. G.H. Sigel, Jr., and D.C. Tran, "Ultra-low loss optical fibers: an overview," *SPIE Proceedings* **484** (1984) pp. 2–6.

29. G. Lu and I. Aggarawal, "Recent advances in fluoride glass fiber optics in the USA," Fourth International Symposium on Halide Glasses, Monterey, California (1987).

30. T. Kanamori, "Recent advances in fluoride glass fiber optics in Japan," Fourth International Symposium on Halide Glasses, Monterey, California (1987).

31. P.W. France, S.F. Carter, M.W. Moore, and J.R. Williams, "Optical loss mechanisms in ZrF_4 glasses and fibres," Fourth International Symposium on Halide Glasses, Monterey, California (1987).

32. P.W. France, J. Williams, S.F. Carter, and K.J. Beales, "Mechanical properties of IR transmitting fibres," paper 11, Second Int. Symposium on Halide Glasses, Troy, NY (1983).

33. J.J. Mecholsky et al., "Fracture analysis of fluoride glass fibers," paper 32, Second Int. Symposium on Halide Glasses, New York (1983).

34. C.J. Simmons, S.A. Azali, and J.H. Simmons, "Chemical durability studies of heavy metal fluoride glasses," paper 47, Second Int. Symposium on Halide Glasses, New York (1983).

35. P.C. Schultz, L.J.B. Vacha, C.T. Moynihan, B.B. Harbison, K. Cadien, and R. Mossadegh, "Hermetic coatings for bulk fluoride glasses and fibers," Fourth International Symposium on Halide Glasses, Monterey, California (1987).

2.7.2 Subsection 2.2

1. M.J. Brau et al., U.S. Patent Number 3,360,649 (1967).

2. A.R. Hilton, Summary Technical Report for New High Temperature Infrared Transmitting Glasses, ONR Report Number 447972 (1962–1964).

3. A.R. Hilton, C.E. Jones, M. Brau, *Infrared Physics* **6** (1966), pp. 183–194.

4. A.R. Hilton, C.E. Jones, M. Brau, *Infrared Physics* **4** (1964), pp. 213–221.

5. A.R. Hilton, D.J. Hayes, and M.D. Rechtin, *Journal of Non-Crystalline Solids* **17** (1975), pp. 319–338.

6. B.R. Kettlewell, B.E. Kinsman, A.R. Wilson, A.M. Pitt, J.A. Savage, and P.J. Webber, *Journal of Materials Science* **12** (1977), pp. 451–458.

7. T. Katsuyama, K. Ishida, S. Satoh, and H. Matsumura, *Appl. Phys. Lett.* **45**:9 (1984), pp. 925–927.

8. J.A. Savage and S. Nielsen, *Infrared Physics* **5** (1965), pp. 195–204.

9. E. Hartouni, F. Hulderman, and T. Guiton, *SPIE Proceedings* **505** (1984), pp. 131–140.

10. P. Klocek and L. Colombo, "Index of refraction, dispersion, band gap and light scattering in GeSe and GeSbSe glasses" *Journal of Non-Crystalline Solids* **93** (1987), pp. 1–16, and "Multiphonon absorption in GeSe and GeSbSe glasses," submitted to *Journal of Non-Crystalline Solids* (1989).

11. J.A. Muir and R.J. Cashman, *Journal of Optical Society of America* **57** (1967), pp. 1–3.

12. S. Shibata, Y. Terunuma, and T. Manabe, "Sulfide glass fibers for infrared transmission," *Mater. Res. Bull.* **16** (1981), pp. 703–714.

13. P.J. Webber and J.A. Savage, *Journal of Non-Crystalline Solids* **20** (1976), pp. 271–279.

14. V.G. Plotnichenko and V.K. Sysoev, *Soviet Journal of Quantum Electronics* **14** (1984), pp. 133–134.

15. P. Klocek, M. Roth, and D. Rock, *Optical Engineering* **26**:2 (1987), pp. 88–95.

16. N.I. Voronin et al., *Sov. Phys. Dokl.* **30**:4 (1985), pp. 306–307.

17. T. Katsuyama, S. Satoh, and H. Matsumura, *J. Appl. Phys.* **59**:5 (1986), pp. 1446–1449.

18. J.I.B. Wilson and D. Blanc, *SPIE Proceedings* **618** (1986), pp. 118–123.

19. T. Miyashita and T. Manabe, *IEEE Transactions Microwave Theory and Techniques*, **MTT-30** (1982), pp. 1420–1438.

20. A. Bornstein, N. Croitoru, and E. Marom, "Chalcogenide infrared glass fibers," *SPIE Proceedings* **320** (1982), pp. 102–105.

21. S. Shibata, M. Horiguchi, K. Kinguji, S. Mitachi, K. Kanamori, and T. Manabe, "Prediction of loss minima in infrared optical fibers," *Electronics Lett.* **17** (October 1981), pp. 775–777.

22. A.M. Bagrov, P.I. Baikalov, A.V. Vasil'ev, G.G. Devyatykh, E.M. Dianov, V.G. Plotnichenko, I.V. Skripachev, V.K. Sysoev, and M.F. Churbanov, *Soviet Journal of Quantum Electronics* **13** (1983), pp. 1264–1266.

23. M.E. Lines, *J. Appl. Phys.* **55**:1 (1984), pp. 4058–4063.

24. T. Kanamori, Y. Terunuma, S. Takahashi, and T. Miyashita, *Journal of Lightwave Technology* **LT-2** (1984), pp. 607–612.

25. G.G. Devyatykh and E.M. Dianov, *SPIE Proceedings* **484** (1984), pp. 105–109.

26. T. Katsuyama and H. Matsumura, *Appl. Phys. Lett.* **49**:1 (1986), pp. 22–23.

27. A.M. Bagrov et al., *Soviet Journal of Quantum Electronics* **15**:10 (1985), pp. 1427–1429.

28. P. Klocek, R. Beni, J. O'Connell, and C. Van Vloten, *SPIE Proceeding* **618** (1986).

29. D.L. Wood and J. Tauc, *Phys. Rev. B* **5**:8 (1972), pp. 3144–3151.

30. A. Bornstein, N. Croitoru, E. Marom, *SPIE Proceedings* **484** (1984), pp. 9–104.

31. T. Hattori, S. Sato, T. Fujioka, S. Takahashi, and T. Kanamori, *Electron. Lett.* **20**:811 (1984).

32. S. Sato, S. Watanabe, T. Fujioka, M. Saito, and S. Sakuragi, *Appl. Phys. Lett.* **48**:15 (1986), pp. 960–962.

33. D.J. McEnroe, M.J. Finney, P.H. Prideaux, P.C. Schultz, *SPIE Proceedings* **799** (1987), pp. 39–43.

34. N.J. Pitt and M.G. Scott, *Guided Optical Structures in the Military Environment*, AGARD–CD–383 (May 1986), pp. 25-1–25-8.

35. M. Saito, M. Takizawa, M. Miyagi, *Journal of Lightwave Technology* **6**:2 (1988), pp. 233–239.

36. N.S. Kapany and R.J. Simms, "Recent developments of infrared fiber optics," *Infrared Phys.* **5** (1965), pp. 69–80; also in N.S. Kapany, *Fiber Optics, Principles and Applications*, Academic Press, New York (1967).

37. M. Saito, M. Takizawa, S. Sakuragi, and F. Tanei, *Applied Optics* **24** (1985), pp. 2304–2308.

2.7.3 Subsection 2.3

1. F. Rosenberger, *Fundamentals of Crystal Growth I*, Springer–Verlag, New York (1979).

2. *The Infrared Handbook*, ed. by W.L. Wolfe and G.J. Zissis, Office of Naval Research, Department of the Navy (Arlington, Virginia, 1978).

3. P. Klocek, ed., *Handbook of Infrared Optical Materials,* Marcel Dekker, New York, in press.

4. J.A. Harrington, D.M. Henderson, A. Standlee, and R.R. Turk, *Infrared Fiber Optics*, Final Technical Report. RADS–TR–80–290, Rome Air Development Center, Griffiss Air Force Base, New York (1980).

5. T.J. Bridges, J.S. Hasiak, and A.R. Strnad, *Optics Letters* **5**:3 (1980), pp. 85–86.

6. R.S. Feigelson, W.L. Kway, and R.K. Route, *SPIE Proceedings* **484** (1984), pp. 133–142.

7. L.G. DeShazer and S.C. Rand, *SPIE Proceedings* **618** (1986), pp. 95–102.

8. M. Kimura, S. Kachi, and K. Shiroyama, *SPIE Proceedings* **618** (1986), pp. 85–88.

9. Y. Mimura, Y. Okamura, and C. Ota, *J. Appl. Phys.* **53** (1982), pp. 5491–5497.

10. G.A. Magel, D.H. Jundt, M.M. Fejer, and R.L. Byer, *SPIE Proceedings* **618** (1986), pp. 89–94.

11. R.S. Feigelson, "Preparation of Infrared Optic Fibers Using New Materials," NRL Number A132311, Contract Number N00014-82-K-2001 (1983).

12. J.A. Harrington, A.G. Standlee, A.C. Pastor, and L.G. DeShazer, *SPIE Proceedings* **484** (1984), pp. 124–127.

13. S. Kachi, K. Nakamura, M. Kimura, and K. Shiroyama, *SPIE Proceedings* **484** (1984), pp. 128–132.

14. J.A. Harrington, *SPIE Proceeding* **266** (1981), pp. 10–15.

15. A.L. Gentile, M. Braunstein, D.A. Pinnow, J.A. Harrington, D.M. Henderson, L.M. Hobrock, J. Myer, R.C. Pastor, and R.R. Turk, "Infrared Fiber Optical Materials," in *Fiber Optics: Advances in Research and Development*, ed. by B. Bendow and S.S. Mitra, Plenum Publishing, NY (1979), pp. 105–118.

16. K. Nassau, *SPIE Proceedings* **320** (1982), pp. 43–48.

17. G.G. Devyatykh and E.M. Dianov, *SPIE Proceedings* **484** (1984), pp. 105–109.

18. D.A. Pinnow, A.L. Gentile, A.G. Standlee, A.J. Timper, and L.M. Hobrock, "Polycrystalline fiber optical waveguides for infrared transmission," *Appl. Phys. Lett.* **33** (1978), pp. 28-29.

19. S. Sakuragi, *SPIE Proceedings* **320** (1982), pp. 2–9.

20. Y. Okamura, Y. Mimura, Y. Komazawa, and C. Ota, "CsI crystalline fiber for infrared transmission," *Japan. J. Appl. Phys.* **19**:10 (October 1980), pp. L649–L651.

21. D. Chen, R. Skogman, E. Bernal G., and C. Butter, "Fabrication of Silver Halide Fibers by Extrusion," *Fiber Optics: Advances in Research and Development*, ed. by B. Bendow and S.S. Mitra, Plenum Publishing, NY (1979), pp. 119–122.

22. K. Takahashi, N. Yoshida, and K. Yamauchi, *Sumitano Electric Technical Review* **26** (1987), pp. 102–109.

23. J.A. Harrington and M. Sparks, *Optics Letters* **8**:4 (1983), pp. 223–225.

2.7.4 Subsection 2.4

1. D.L. Wood, K. Nassau, and D.L. Chadwick, *Applied Optics* **21** (1982), pp. 4276–4279.

2. H. Takahashi, I. Sugimoto, T. Sato, and S. Yoshida, "GeO_2-Sb_2O_3 glass optical fibers for 2–3 μm fabricated by VAD method," *SPIE Proceedings* **320** (1982), pp. 82–92.

2.7.5 Subsection 2.5

1. T. Hidaka, T. Morikawa, and J. Shimada, *J. Appl. Phys.* **52**:7 (1981), pp. 4467–4471.

2. E. Garmire, *SPIE Proceedings* **320** (1982), pp. 70–78.

3. M. Miyagi, K. Harada, Y. Aizawa, and S. Kawakami, *SPIE Proceedings* **484** (1984), pp. 117–123.

4. E. Garmire, *SPIE Proceedings* **484** (1984), pp. 112–116.

5. C.A. Worrell, *SPIE Proceedings* **843** (1987), pp. 80–87.

2.7.6 Subsection 2.6

1. R. Altkorn, M.E. Marhic, and F. Tawg, *SPIE Proceedings* **843** (1987), pp. 130–136.

2. L. Stone, *IEEE J. Quantum Electron.* **8** (1972), pp. 386–388.

Section 3
Applications of IR Fiber Optics

3.1 OVERVIEW

Silicon-based optical waveguides function quite well to wavelengths of approximately 2 μm, demonstrating peak transparency in the near-IR region of 1.3 to 1.6 μm and approaching 0.15 dB/km at the latter wavelength. The need for optical fibers with IR transmission extended beyond that of silicon is driven by a variety of applications that will be briefly summarized in this overview. Since each material's class of IR fibers tends to focus on or be associated with specific applications, the general introduction is followed by a detailed treatment that highlights applications for fluoride glass, chalcogenide glass, and crystalline and hollow waveguide configurations. Regardless of the specific type of fiber under consideration, the questions that must always be addressed relate to whether the application demands an IR-transmitting fiber and, if so, what specifications the waveguide must meet optically, mechanically, and environmentally. These specifications will determine which IR fiber described in Sections 1 and 2 could be used.

Many applications of IR fiber optics are being realized today; others, beyond current technology, spur the IR fiber optics for future applications.

3.1.1 General Areas of Application for IR Fibers

Many IR materials, including glasses and crystals, possess intrinsic minimum attenuation levels that are much lower than those of silicon. Applications such as long-distance, repeaterless telecommunication links and high-power laser propagation depend heavily on such low losses. In addition to the ultra-low-loss properties attractive to telecommunications, IR fibers can exhibit very low materials dispersion over a broad spectral range, making them useful for color-multiplexing of high-bandwidth systems. Table 32 provides a list of some of the prospective applications that have been proposed for IR fibers.

3.1.2 IR Laser Propagation

Given the availability of important long-wavelength laser sources such as HF (2.8 μm), DF (3.8 μm), CO (5.0 μm), and CO_2 (10.6 μm), it is desirable to provide fiber waveguides to interface with each of these in the extended IR region. Numerous applications relating to biomedicine, remote laser processing or welding, surface modification, optical powering of electronics and optical radars, to name a few, relate to the use of IR waveguides operating at discrete laser wavelengths. Parameters of interest include measurement of maximum power densities for both steady-state and pulsed-power propagation, damage mechanisms and thresholds in waveguides versus those in bulk materials, thermal loading effects, nonlinear power attenuation mechanisms, interface degradation, and the differentiation of extrinsic and intrinsic waveguide limitations.

TABLE 32. APPLICATIONS OF INFRARED FIBERS

- Long-distance repeaterless data links
- Optical power transmission
- Remoting of focal plane arrays
- Fiber optic shutters
- Laser annealing and processing
- On-line pollution and gas analysis
- Low-noise fiber sensors
- Radiation-resistant data links
- IR imaging
- Remote IR spectroscopy
- Fiber optic lasers
- IR delay lines
- Medical surgery and cauterization
- Temperature monitoring
- High-bandwidth color multiplexing
- Remote powering of electronic instrumentation
- Microwelding and cutting
- Nonlinear components

3.1.3 IR Imaging

The extended spectral transparency of halide and chalcogenide materials relative to oxides is significant for IR imaging applications. Atmospheric windows in the 3- to 5-μm and 8- to 14-μm range make these wavelengths important for both civilian and military avionic systems. IR fibers permit remoting of focal plane arrays for sensing applications. Coherent bundles of fibers permit high resolution, remote use, thermal imaging, and tomography. In certain materials such as the fluorides, both visible and IR imaging might be possible via the same waveguide. Remote thermography is of interest for real-time diagnostics of many industrial processing operations or for monitoring such things as engine performance and thermal stability.

3.1.4 Active Components-Fiber Lasers

While IR materials can be used in waveguide configurations for propagation of IR laser radiation, they may also serve as the media for generating the light itself. The heavy metal fluoride glasses, for example, have been shown to be excellent hosts for a variety of rare-earth elements that possess numerous emission lines in the middle IR region. Laser applications include fiber optic sources for IR links, excitation sources for fluorescent fiber sensors, atmospheric pollution monitoring similar to the role played by lead salt laser's, and optical radar or ranging systems. Questions of interest here include measurement of relevant parameters such as quantum efficiency, lifetime, spectral output, optimum doping level, optimum host glass composition, and the effects of aging on laser performance.

3.1.5 Sensors

A major applications area yet to be exploited for IR fiber optics relates to that of sensors. Direct, remote temperature measurements of passively radiating objects is perhaps the simplest and most obvious sensor opportunity offered by IR waveguides. However, a transparent host waveguide in the IR offers the possibility of designing a large number of chemical sensors based on evanescent absorption at the core-cladding interface. Reactive coatings, porous coatings, and photoresponsive or fluorescent coatings are of potential interest. Remote sensing over long distances becomes possible because of low losses. Selective incorporation of ions such as manganese into fluoride glasses permits the fabrication of magnetically sensitive fibers. Very low Rayleigh backscatter coupled with low optical losses may offer some advantages in fiber gyro sensor design. Much of this technology clearly lies in the future.

3.1.6 Radiation-Hardened Links

Most optical materials, including presently available silicon and doped silicas employed for optical fibers, suffer optical degradation when exposed to ionizing radiation. The optical absorption arising from defects or dangling bonds in the glass matrix is typically most intense in the near-ultraviolet and visible regions of the spectrum with long tails trailing off into the infrared. There is some experimental evidence that IR fibers may be less susceptible to damage arising from nuclear radiation and, having lower melting points, may recover more rapidly at room temperature if damage does occur. More testing is needed to verify and examine the nature of the radiation-induced optical damage in IR waveguides. However, certain selected fluoride glass compositions appear promising.

3.1.7 Medical Uses

The use of IR fibers for medical applications should prove to be exciting and important. The energy deposition of long wavelength light such as 10.6 μm from CO_2 and 5 μm from CO lasers, for example, is extremely localized. Cancerous tumors can be destroyed without harming adjacent healthy tissue. However, fiber waveguides that are both small and flexible and, if possible, nontoxic are needed for use within the body. KRS–5 fibers have been used primarily for CO_2 lasers, but chalcogenides of certain compositions appear to offer some promise. Lasers and IR fibers can be used for remote cauterization of bleeding ulcers, to remove plaque from the wall of blood vessels, and for laser drilling of teeth. Internal imaging of organs or scanning of body areas can often point to the presence of tumors. Fiber-optic biomedical and biochemical sensors may permit the on-line monitoring of drugs or medications administered to patients, as well as providing information on local blood pressure, temperature, velocity or oxygen level. Critical issues here include selection of waveguide materials and compatible coatings that provide chemical durability and mechanical strength.

3.1.8 Nonlinear Optics

Finally, IR fibers, especially single-crystal waveguides, potentially offer interesting applications for nonlinear optics. Fibers provide a much longer optical pathlength than conventional bulk materials for generating stimulated Raman and Brillouin scattering. Second harmonic generation and differential frequency mixing could be used to achieve tunable light sources. Optical switching and power-dependent real-time optical filters can also be fabricated from IR fibers.

Numerous other applications for IR fibers probably exist. This brief overview has shown that current silica fiber technology falls short in many areas. Perhaps the user wants greater transparency or range, the transfer of an IR image, or the propagation of a CO_2 laser beam. These and many other potential applications are the stimuli for developing both present and future IR fibers. For the most part, IR fibers are just becoming available for commercial use. The fibers are crude compared with intrinsically possible waveguides, but they represent a beginning.

The following subsections present a detailed discussion and analysis for some of the selected applications of IR waveguides appropriate for each material type or class. The first class to be considered are the fluorides, which offer perhaps the most intrinsically transparent candidate materials for IR fibers but which cannot be used for longer wavelength applications (beyond about 8 μm), depending on both the fiber length and the specific fluoride composition. Without question, however, the major application of interest for fluoride fibers is long-distance, repeaterless data transmission.

3.2 TELECOMMUNICATIONS

As discussed earlier in this book, fluoride fibers can possess intrinsic attenuation levels approaching 10^{-3} dB/km. Assuming source and detector performance similar to that available in the near IR, these loss values translate into repeaterless operation over spans more than 1,000 km long, perhaps as long as 10,000 km. Transoceanic links are attractive applications for such fibers since repeater powering, maintenance, and repair present problems in undersea applications. Bendow[1] has reviewed the most promising components for such an undersea link (Table 33).

In general, it is useful to consider the source and detector options in the construction of the link. For fluorides, it is necessary to focus primarily in the 2- to 5-μm range. Table 34 contains a listing of possible semiconductor lasers (after Reference 1).

At shorter wavelengths, and for laboratory use, color center lasers are attractive as light sources for fiber links. Here, an external modulator is used. Some color center lasers are listed in Table 35 (after Reference 1).

Photodetectors seem to be less critical than light sources in the 2- to 5-μm range. Table 36 lists several of these candidates that might be used in an IR link (after Reference 1).

Recently, external modulators based on $LiNbO_3$ have been fabricated for use at 2.6 μm and 3.3 μm for the purpose of demonstrating an IR data link.

Low loss also is attractive for IR optical power transfer. Nippon Telegraph and Telephone has used a Gd-Ba-Zr-Al fluoride glass fiber in conjunction with an HF laser operating at 2.7 μm. Typical delivered power was on the order of 1 watt for a 300-μm core fiber about 4 feet long. The basic experimental setup is shown in Figure 102. At higher power levels, the input face of the fluoride fiber was observed to melt. Unfortunately, there is a strong OH absorption in the fluoride fiber at this wavelength (2.7 μm) so that internal heat is expected to be strong. Ultimately, power transfer at wavelengths near 2.5 μm and above 3.7 μm may be feasible in the fluoride fibers.

TABLE 33. MOST PROMISING CANDIDATES FOR ULTRALONG LINK, UNDERWATER MID-IR FIBER OPTICS SYSTEMS

Component	Most Promising Candidate	Fabrication Method	Other Attributes	Backup Choice
Fibers	HMF glass	Furnace-drawn from preform	Must be hermetically coated and plastic jacketed	Chalcogenide glass
Light sources	Semiconductor lasers (III–V alloys)	VPE, MBE, or MO-CVD	Cooling may be required	IV–VI SCLs
Detectors	HgCdTe or III–V alloys	VPE, MBE, or MO-CVD	Cooling will be required	Doped semiconductors or Schottky barrier detectors

TABLE 34. PROPOSED 2- TO 5-μm SEMICONDUCTOR LASERS: IMPORTANT FEATURES

Material Type	Wavelength (Tunability)	Substrates	Composition Grading	Growth Technique	Temperature
(GaIn)Sb/(AlIn)Sb	1.8 to 5.0 μm preset by composition (no)	GaAs	Yes, $Ga_{1-x}In_xSb$	MBE	77°K (maximum ≤100°K)
InGaAsSb	To 3.8 μm preset (no)	InAs, InSb, GaSb	Yes (<3.8)	MOCVD	Unspecified
InGaAsPb	1.2 to 1.6 μm preset (no)	InP	No		Thermoelectric cooler at room temperature
PbCdS or PbSSe	2.5 to 4.35 μm 4.35 to 8.5 μm (yes, via temperature)	PbTe, PbSe, PbS	No		Cryotemp 20° to 60°K
InAsP/InGaAs	1.7 to 3.5 μm preset (no)	InP, InAs	Yes	VPE	Unspecified

TABLE 35. PERFORMANCE OF COLOR CENTER LASERS

Host	LiF	NaF	KF	NaCl	KCl:Na	KCl:Li	KCl:Na	KCl:Li	RbCl:Li
Center (μm)	F_2^+ 0.647	$(F_2^+)^*$ 0.870	F_2^+ 1.064	F_2^+ 1.064	$(F_2^+)_A$ 1.340	F_B(II) 1.340	F_A(II) 0.647	F_A(II) 0.647	F_A(II) 0.647
Pump power	4.00 W	1.00 W	5.00 W	5.00 W	0.10 W	0.15 W	1.50 W	2.60 W	2.20 W
Tuning range (μm)	0.82 1.05	0.99 1.22	1.22 1.50	1.40 1.75	1.62 1.91	2.00 2.50	2.25 2.65	2.50 2.90	2.60 3.30
Maximum power output (cw)	1.8 W	400.0 mW	2.7 W	1.0 W	12.0 mW	10.0 mW	35.0 mW	240.0 mW	55.0 mW

TABLE 36. POTENTIAL DETECTORS FOR MID-IR FIBER OPTICS

Detector Type	Comments
HgCdTe	All-around good candidate: high sensitivity, high-frequency response, well-established fabrication technology
Lead Salts	Presently lower performance than HgCdTe, but might be improved
III-V Semiconductor Alloys	Could be matched to SCLs to provide high sensitivity and frequency response; none currently in existence
Doped Semiconductors	Extrinsic Si and Ge are potential candidates, but more work is required
Metal Silicide Schottky Barrier Detectors	Good frequency response, but requires increased sensitivity; well-established fabrication technology

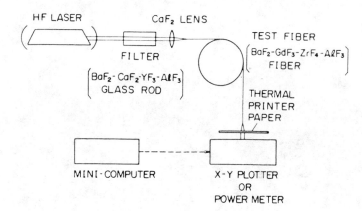

Figure 102. Experimental Arrangement for Measuring 2.7-μm Power Transmission Properties (X-Y Plotter and Minicomputer Were Used for Laser Printer) (Reference 2)

3.3 IR FIBER BUNDLES APPLICATIONS

3.3.1 Image Bundle

IR fiber optic bundles can be formatted for coherent or incoherent applications. An incoherent IR fiber bundle would be used to flexibly transfer energy from one point to another with no regard for spatial resolution. A coherent bundle is generally used where spatial resolution is important, as in thermal imaging, where image plane transfer is required. Generally, coherent bundle applications include remotely locating an image plane or detector array, combining detector arrays and image manipulation. Whether coherent or incoherent, the transmission through a bundle is not simply the fiber attenuation but also the Fresnel reflection loss at both ends of the bundle and loss owing to the packing fraction. Since the active area of a fiber is the core only, bundles with hexagonal packing of cylindrical fibers have a packing fraction of $\pi/2 \sqrt{3} \, d^2/D^2$, where d = core diameter and D = overall fiber diameter, and square fibers packed hexagonally have a packing fraction of d^2/D^2.

It has been shown that IR fiber optic image bundles can be used with IR detector arrays and maintain reasonable system performance if the strict design criteria in Table 37 are met.[3]

TABLE 37. CRITERIA FOR FIBER OPTIC IMAGE BUNDLE
[Assumes IR detector array, 65-μm center-to-center spacing and 85-percent active (55 μm)]

Requirement	Fiber Specifications
Minimum crosstalk less than 5 percent (means that percentage of energy coupling into an adjacent fiber), maximizes system MTF and performance	• Fused length of fiber in bundle must be less than 5 cm • Core-to-fiber diameter ratio must be less than 0.8 • 1:1 coupling of fiber and detector pixel
Maximize bundle transmission (maximize packing fraction)	Large core-to-fiber diameter ratio (0.75 to 0.8 near optimum)
Maximize matching or alignment between each fiber and detector active area	Fractional shift of fiber core center with detector active area center must be less than 10 percent

Sample applications of IR coherent fiber optic image bundles include: thermal imaging in inaccessible areas, owing to the flexibility of the fiber bundle; full 360-degree fields of view, with a single central image processor, having multiple inputs from several strategically positioned fiber bundles; relocation of a detector array allowing system configuration and packaging advantages; increasing the field of view of a thermal imaging system by combining more than one detector array together with a split fiber bundle, allowing a greater field of view while maintaining the resolution of the system. Figure 103 shows some of these coherent IR image bundle configurations. Figures 104 through 107 show some examples of IR image bundle applications.

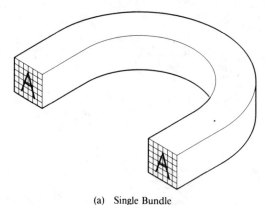

(a) Single Bundle

Note: Allows Remote Location Of Focal-Plane Array Or Detectors Of Any Configuration.

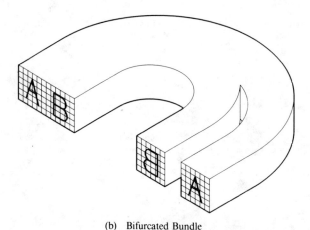

(b) Bifurcated Bundle

Note: IR Fiber Optic Bifurcated Image Bundle Combines Focal-Plane Arrays which Increases Field of View While Maintaining Resolution.

Figure 103. Coherent IR Fiber Optic Bundles (Reference 3)

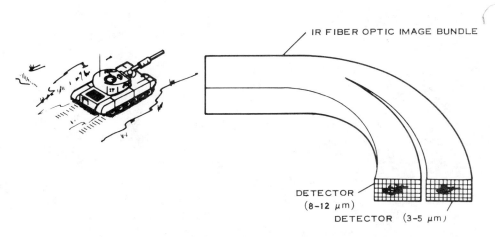

Figure 104. IR Fiber Optic Bundle Allowing Simultaneous Imaging
at Two Different Wavelength Regions (Reference 3)

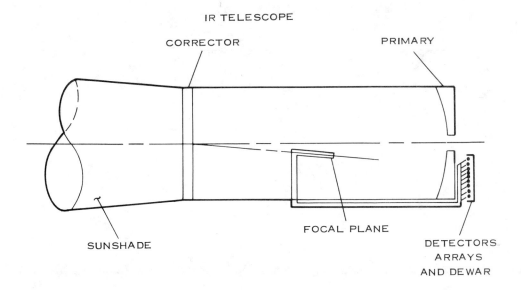

Figure 105. Flexible Coherent Image Bundle (Detectors and Cryogenics Can Be Remoted,
Allowing More Unobstructed Image) (Reference 4)

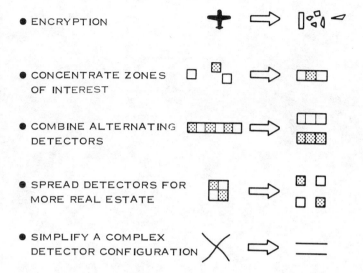

Figure 106. IR Image Bundle Configurations (Reference 5)

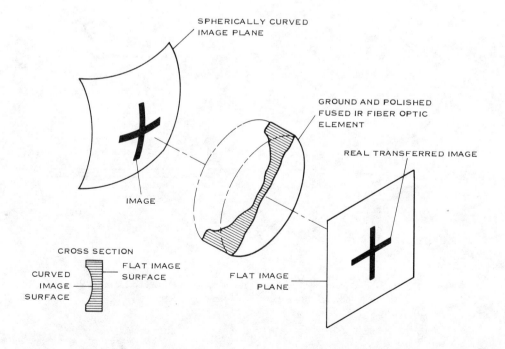

Figure 107. IR Fiber Bundle, Field Flattener (Reference 6)

3.3.2 Tapered Bundle

A tapered bundle, as shown in Figure 108, increases the light flux density at the fiber exit face compared with the fiber entrance face. Accompanying this change is a proportionate increase in solid angle. The equation relating the effects of a tapered bundle is:

$$\frac{d_1}{d_2} = \frac{NA_2}{NA_1} \tag{1}$$

where

d_1 = entrance face fiber diameter

d_2 = exit face fiber diameter

NA_1 = numerical aperture of entrance face

NA_2 = numerical aperture of exit face

$\frac{d_1}{d_2}$ = taper ratio.

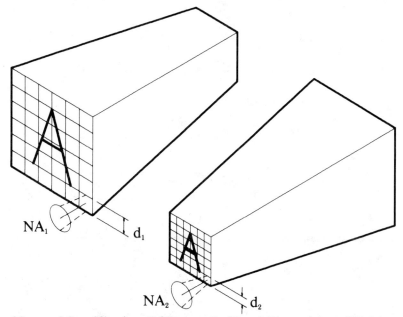

Note: Magnification Of Image Or Focal-Plane Array, Which Increases The Effective Focal Length Of The System While Maintaining Resolution And Field Of View

Figure 108. Taper IR Fiber Bundle (Reference 3)

If the input energy on the entrance of the tapered bundle falls within the entrance NA as defined by Equation (1), the efficiency of the energy transfer to the exit face in an ideal taper with no losses is 100 percent. This efficiency decreases if the entrance NA is overfilled due to the leakage energy out of the tapered waveguide since it encounters the core/clad interface at angles greater than the critical angle for total internal reflection. The energy density gain is simply $(d_1/d_2)^2$. One application of a tapered fiber bundle is as a relay lens system because of its ability to function in a fast optical system [small f/number; f/no. = 1/(2 NA)].

IR tapered bundles can be used in thermal imaging or sensing systems that have a fixed field of view but require an increased sensitivity or range or vice versa. While this can be done with IR relay lenses whose magnification would increase the energy density on the detector array, relay lenses are difficult to make with small f/numbers. Therefore, greater magnifications can be achieved with the fiber taper than with a relay lens in a fast system. In slower systems where relay lenses are effective, the small size of the fiber taper might offer packaging advantages.

A tapered IR fiber bundle can also be configured with different taper ratios in different segments of the bundle so that magnification is not uniform as shown in Figure 109. This has applications in thermal imaging systems where different segments of the image must be emphasized.

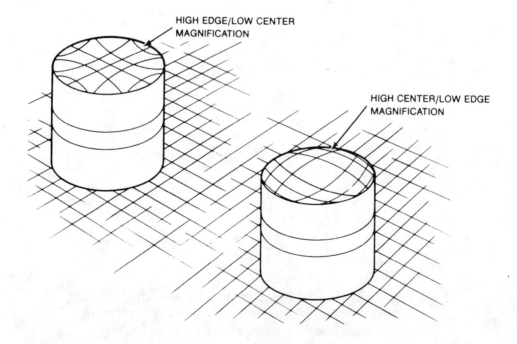

Figure 109. Nonuniform Magnifying IR Bundle (Reference 6)

3.3.3 IR Fiber Optic Reformatter

An IR fiber optic reformatter (Figure 110), changes the field-of-view geometry at the entrance face to another at the exit face. It can replace an anamorphic relay lens system and eliminate spatial dependence of the system speed inherent in relay lenses as well as being physically smaller. This independence between the system speed and the field of view provided by the fiber reformatter could increase flexibility in optical design and packaging in an IR system.

An example of an IR fiber reformatter application is shown in Figure 111 where a spot-to-line reformatter is used instead of a slit for a narrow, square field-of-view grating spectrometer. This shrinks the size of the spectrometer by eliminating the large slit while maintaining the field of view now defined by the entrance face (square spot) of the reformatter while the exit face (line) produces the slit image.

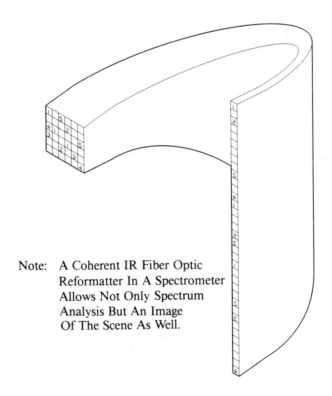

Note: A Coherent IR Fiber Optic Reformatter In A Spectrometer Allows Not Only Spectrum Analysis But An Image Of The Scene As Well.

Figure 110. IR Fiber Optic Image Reformation (Coherent) (Reference 3)

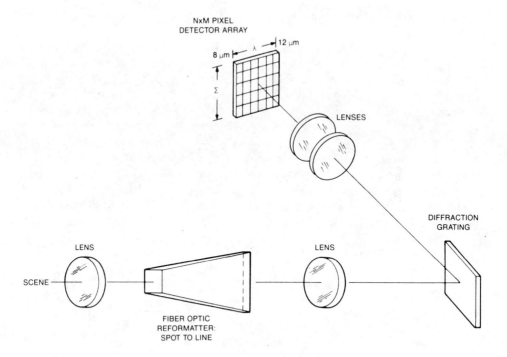

Figure 111. Operation of Spectrometer Having IR Fiber Optic Reformatter (Reference 4)

3.4 SINGLE IR FIBER APPLICATIONS

Single IR fiber applications[3,5] that enhance existing systems and provide novel system concepts are numerous. Remotely locating detectors, sources, and temperature and pressure sensors are some of the general applications discussed below. Subsection 3.1.1 has described many of these applications.

3.4.1 Temperature Sensor

A temperature sensor[3] designed with flexible IR optical fibers could operate over a wide temperature range as low as room temperature or below and be very useful for electronics troubleshooting, process control, and medical diagnosis. The temperature sensor design shown in Figure 112 simultaneously detects two wavelength regions. The ratio of the intensity of these two spectral bands determined by two detectors, yields the temperature of the viewed object.

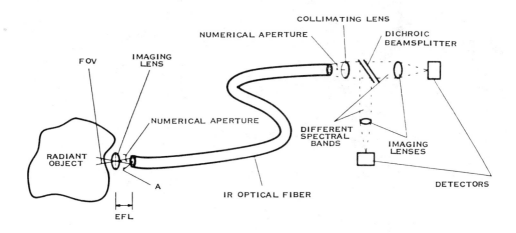

Figure 112. Temperature Sensor Diagram (Reference 3)

A signal-to-noise ratio was calculated[7] by:

$$SNR = W * K * \frac{(EFL)^2 \, NA \, \tau_o}{\sqrt{A}} \qquad (2)$$

where

SNR = sensor threshold signal-to-noise ratio
W = incident energy on system
K = sensor parametric constant
A = area of fiber
EFL = effective focal length of sensor
NA = numerical aperture of fiber
τ_o = optical transmission of fiber and system

for a sample system with a pyroelectric detector looking at a 300°K blackbody in the 3- to 4-μm and 10- to 12-μm spectral passbands so that

SNR = 6
A = 300-μm² fiber
EFL = 0.5 cm
NA = 0.5
τ_o = 70 percent

and the ratios of the two spectral passbands are 63 at 300°K and 47 at 310°K. This indicates that a temperature measurement with accuracies of 1°C is possible.

3.4.2 Pressure Sensor

IR fiber optic interferometric, as diagrammed in Figure 113, acoustic sensors could offer greater sensitivity over typical single-mode silica optical fibers with coatings that raise the pressure sensitivity of the fiber. The optical phase retardation of the sensing fiber is given by:[8]

$$\frac{\Delta \phi}{\phi} = \epsilon_z - \frac{n^2}{2} [(P_{11} + P_{12}) \epsilon_r + P_{12} \epsilon_z] \qquad (3)$$

where

P_{11} and P_{12} = elastoptic coefficients of the fiber core
n = index of refraction of the core
ϵ_z and ϵ_r = strain components.

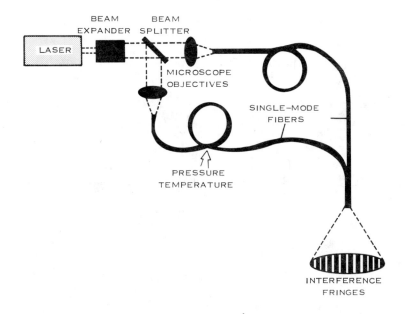

Figure 113. Simplified Diagram of Pressure Sensor (Reference 5)

If the physical properties of silica and GeSe glasses are compared in the above equation, the sensitivity of the chalcogenide fiber is nearly an order of magnitude higher than that of the silica, mainly because of the higher indices.[3] More sensitive acoustic sensors could, therefore, be designed with chalcogenide glass optical fibers.

3.4.3 Remote Location of Detector or Source

There are many applications of IR optical fibers that involve remoting the source(s) or detector(s). One example of these applications is a 360-degree field of regard sensor or detector with several IR fibers being routed to a central detector or array as shown in Figure 114.

Probably the most exciting remote source application involves the flexible transmission of laser energy. The use of lasers in medicine for surgery and other procedures is an expanding field. The potential for selective and highly confined tissue cutting using lasers at approximately 100 watts or less is the driving force for much medical research and the development of IR fiber optics. Three-dimensional flexible delivery of laser energy would greatly extend the application of lasers in medicine. Various lasers are of interest for medical applications, but the CO_2 and CO lasers that operate at 10.6 and 5 μm, respectively, are those for which the IR optical fibers discussed in this text are best suited. For CO_2, the solid dielectric fibers made of the halide crystals or chalcogenide glasses appear to be the best choices, while for CO the fluoride glasses are probably best. The reasons for this selection are based on transmission and flexibility. It is the flex-

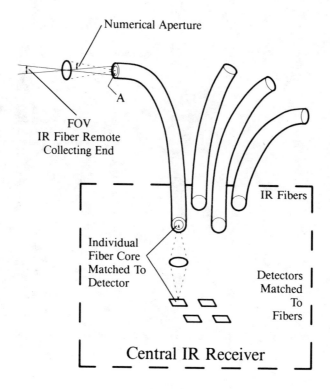

Figure 114. Threat-Warning Receiver Diagram (Reference 3)

ibility and size requirements that limit hollow IR waveguides for many medical applications. Other medical laser delivery system considerations include toxicity, which must either be eliminated by the choice of IR fiber material or be packaged adequately to ameliorate this concern, and optical coupling from the laser to the fiber and the output of the fiber. The specifics of the optics depend on the application in terms of desired modes, focal length, f-number, and spot size. In general, however, the lens system f-number should be matched to the fiber NA using standard laser optical components.

Other remote laser source aplications include remote optical powering. Here, laser energy is delivered via the fiber to the system to be powered where the optical laser energy is converted to available electrical energy. This is of interest in remote systems that currently rely on batteries.

Remote lasers for optical switches or fuses for explosives, laser welding, and laser cutting are other applications of IR fibers. Multiple outputs from a centrally located CO_2 laser could be an interesting rangefinder application as the system field of regard is increased without increasing the range-limiting divergence of the laser.

The hollow waveguides may have a high laser damage threshold. However, absorption, large sizes, and limited flexibility restrict bending, which may limit these interesting waveguides to short and stiff applications, but with the potential for high powers (hundres of watts). One way of increasing the laser power handling capability of the solid dielectric fibers is shown in Figure 115, where a bundle of flexible fibers is used to spread the energy across many fibers.

A remote blackbody source to provide absolute radiometric calibration or dc restoration in a thermal imaging system is useful but difficult to design because of packaging constraints. An IR optical fiber can remote this source and guide the reference into a convenient image plane, thereby overcoming the packaging difficulties. Boresighting for forward-looking infrared imaging (FLIR) systems is a similar application whereby the fiber brings a remote source into the system to check the boresight alignment (Figure 116).

Another remote blackbody source application is a gas analyzer for medical or industrial use. This is shown in Figure 117.

The blackbody IR source is transmitted to a remote gas cell and is collimated. It then traverses the cell where gas species (e.g., CO_2) absorb some of the energy at a characteristic wavelength. The unabsorbed energy is transmitted to a chopper wheel via another fiber. This creates a modulated signal for the detector, which has an output calibrated for the characteristic absorption of different gas species. Other applications of IR optical fibers can be found in References 9 through 20.

144

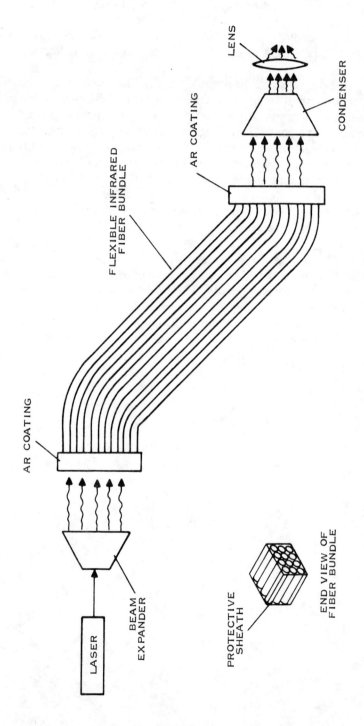

Figure 115. CO_2 Laser Guide (Reference 5)

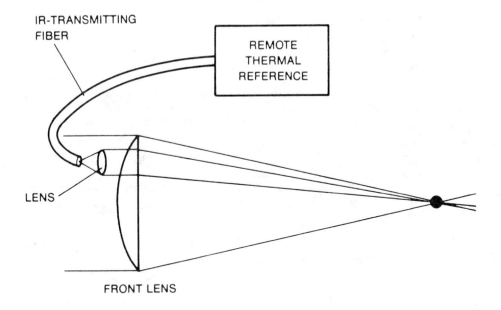

Figure 116. FLIR Boresighting (Reference 4)

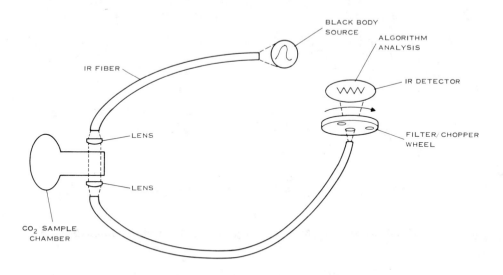

Figure 117. Remote CO_2 Gas Analyzer/Remote Radiometry (Reference 5)

3.5 REFERENCES

1. B. Bendow, *Mid-IR Fiber Optics Technology Study and Assessment*, NOSC Final Report (February 28, 1984).

2. K. Jinguji, M. Horiguchi, S. Mitachi, T. Kanamori, and T. Manabe, "Infrared power delivery in the 2.7 μm band in fluoride glass fiber," *Jpn. J. Appl. Phys.* **20**:6 (1981), pp. L392–L394.

3. P. Klocek, M. Roth, D. Rock, *Optical Engineering* **26**:2 (1987), pp. 88–95.

4. D. Rock, "Development of Chalcogenide Glass Fiber for FLIR Wavelengths," presented at National IRIS Symposium, Gaithersburg, MD (May 1985).

5. P. Klocek and B.M. Kale, "Applications of infrared transmitting glasses in electro-optic systems," presented at the SPIE conference on Infrared Sensor Technology (May 4–5, 1982, Arlington, VA).

6. N.R. Truscott and C.H. Tosswill, *SPIE Proceedings* **306** (1981), pp. 29–39.

7. "The Fundamentals of Thermal Imaging Systems" edited by F. Rosell and G. Harvey. *NRL Report 8311*, Washington, D.C., *EOTPO Report* No. 46, pp. 49–64.

8. N. Lagakos, E.U. Schnaus, J.H. Cole, J. Jarzynski, J.A. Bucaro. *IEEE Journal of Quantum Electronics,* **QE-118**:4 (1982), pp. 683–689.

9. D.A. Pontarelli, "Infrared Fiber Optics," DTIC Report AD–235–192.

10. N.S. Kapany, numerous articles, 1958–1962.

11. A. Katzir, *Laser Focus/Electro Optics* (May 1986), pp. 94–110.

12. A. Katzir, F. Bowman, H. Narciso, Y. Asfeur, A. Zur, *SPIE Proceedings,* **713** (1986), pp. 97–104.

13. E. Sinofsky and G. Gofstein, *AIP Proceedings,* **160** (1987), pp. 710–714.

14. D. Pruss, P. Dreyer, E. Koch, *SPIE Proceedings,* **799** (1987), pp. 117–122.

15. M. Saito, M. Takizawa, K. Ikegawa, H. Takami, *J. Appl. Phys.,* **63**:2 (1988), pp. 269–272.

16. J. Bedrossian, Jr., *SPIE Proceedings* **780** (1987), pp. 90–96.

17. V.N. Perminova and V.K. Sysoev, *Sov. J. Opt. Technol* **52**:4 (1985), pp. 250–251.

18. S. Simhany, E.M. Kosower, A. Katzir, *Appl. Phys. Lett.* **49**:5 (1986), pp. 253–254.

19. A. Zur and A. Katzir, *Appl. Phys. Lett.* **48**:7 (1986), pp. 499–500.

20. S. Simhany and A. Katzir, *Appl. Phys. Lett.*, **47**:12 (1985), pp. 1241–1243.

About the Authors

Paul Klocek joined Texas Instruments in 1984 and is manager of the Advanced Optical Materials Laboratory, where he is responsible for the research and development of optical materials and components. He has been involved in fundamental characterization and materials development of chalcogenide glasses, various III-V and II-VI crystals, diamond, silicon, and various nitrides, from which he has developed optical fibers, windows, domes, and geometric optics. He has chaired SPIE conferences in infrared optical materials and is editor/author for two books and several papers on infrared optical materials and components.

George H. Sigel, Jr. is director of the Rutgers University Fiber Optic Materials Research Program, which is a joint university, industry, government effort directed at broadly based, interdisciplinary research on fiber optic materials, processing, measurements, and devices. Prior to joining Rutgers in 1985, Dr. Sigel was at the Naval Research Laboratory, where he directed fiber optic materials research programs. His research included work on radiation effects on optical fibers, the development of low loss fluoride glass fibers in the IR, and fiber optic sensors. He has published more than 200 papers on optical materials and fibers and is the holder of 14 patents.